1. Jumper the Arduino to the breadboard.

2. Connect the Ping ultrasonic sensor to power, ground, signal.

3. Add the super-bright LEDs and their resistors.

```
void loop() {

    // measure distance: send "Ping"
    pinMode(SensorPin, OUTPUT);
    digitalWrite(SensorPin, HIGH);
    delayMicroseconds(5);
    digitalWrite(SensorPin, LOW);

    // measure distance: listen for "Ping"
    pinMode(SensorPin, INPUT);
    pulseDuration=pulseIn(SensorPin, HIGH);

    // divide by two (back/forth for a single trip), divided by speed of sou
    pulseDuration=pulseDuration/2;
    distance = int(pulseDuration/29);
```

4. Upload the Arduino program to the microcontroller.

5. The LEDs glow to tell distance: blue = far (cold), red = near (hot)!

```
    // divide by two (back/forth for a single trip), divided by speed of sound = di
    pulseDuration=pulseDuration/2;
    distance = int(pulseDuration/29);

    // light up red led: inverted linear of 0-25cm to 0-255 eq. off to max. brightn
    if (distance > 0 && distance < 25) {
      int RedValue=(25-distance)*10.2;
      analogWrite(RedLedPin, RedValue);
    } else {
      analogWrite(RedLedPin, 0);
    }

    // light up blue led: 10-90cm ^= 0-255, 25-50cm ^= 255-0 on BlueLedPin
    if (distance > 10 && distance <= 90) {
      int BlueValue = (distance-10)*17;
      analogWrite(BlueLedPin, BlueValue);
```

6. Experiment with more "hot/cold" Arduino programs.

Learn to use an Arduino to read a distance sensor and control super-bright LED lights, without any soldering!

Parts list:

- USB cable, A to B
- Super-bright Blue LED, 5mm, 30m
- Super-bright Red LED, 5mm, 25m
- Resistor, 56Ω, 1/4W, for Blue LED
- Resistor, 150Ω, 1/4W, for Red LED

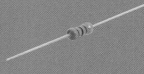  

Arduino Uno microcontroller   **Carbon-film resistor**   **Ping ultrasonic distance sensor**

- Breadboard Jumper Wires

Tools checklist:

- Computer running Arduino software (free download from arduino.cc)
- USB cable, A to B (not pictured)

For complete instructions and details on this project visit:

radioshackdiy.com/hot-cold-LEDs

radioshack™

Columns

Features

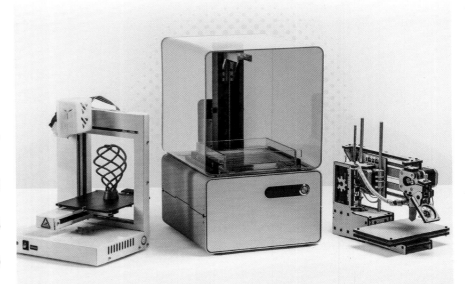

ON THE COVER
Up Plus 2, Form 1, and Printrbot Simple highlight our 3D printer roundup. Spiral Lightbulb Sculpture by benglish.

CONTENTS

ULTIMATE GUIDE TO 3D PRINTING 2014

Projects

Buyer's Guide

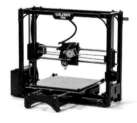
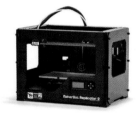

> "If you always want what you can't have, what do you want when you can have anything?"
> —Tagline from Primer, 2004

FOUNDER & PUBLISHER
Dale Dougherty
dale@makezine.com

PRESIDENT & COO
Greg Brandeau
greg@makezine.com

EDITOR-IN-CHIEF
Mark Frauenfelder
markf@makezine.com

VICE PRESIDENT
Sherry Huss
sherry@makezine.com

EDITORIAL

EXECUTIVE EDITOR
Mike Senese
msenese@makezine.com

EDITORIAL DIRECTOR
Ken Denmead
kdenmead@makezine.com

MANAGING EDITOR
Cindy Lum
clum@makezine.com

PROJECTS EDITOR
Keith Hammond
khammond@makezine.com

SENIOR EDITOR
Goli Mohammadi
goli@makezine.com

SENIOR EDITOR
Stett Holbrook
sholbrook@makezine.com

TECHNICAL EDITOR
Sean Michael Ragan
sragan@makezine.com

ASSISTANT EDITOR
Laura Cochrane

STAFF EDITOR
Arwen O'Reilly Griffith

EDITORIAL ASSISTANT
Craig Couden

COPY EDITOR
Laurie Barton

SENIOR EDITOR, BOOKS
Brian Jepson

EDITOR, BOOKS
Patrick DiJusto

CONTRIBUTING EDITORS
William Gurstelle, Charles Platt

CONTRIBUTING WRITERS
John Abella, Tom Burtonwood, James Christianson, Tim Deagan, Stuart Deutsch, Cory Doctorow, Brook Drumm, Anna Kaziunas France, Matt Griffin, Kacie Hultgren, Fred Kahl, Ben Lancaster, Blake Maloof, Chris McCoy, Tyler McNaney, Bob Parks, Derek Poarch, Matt Stultz, Anderson Ta

ONLINE CONTRIBUTORS
Alasdair Allan, John Baichtal, Meg Allan Cole, Michael Colombo, Jimmy DiResta, Nick Normal, Haley Pierson-Cox, Matt Richardson, Andrew Salomone, Andrew Terranova, Glen Whitney

DESIGN, PHOTOGRAPHY & VIDEO

CREATIVE DIRECTOR
Jason Babler
jbabler@makezine.com

ART DIRECTION – SENIOR DESIGNER
Juliann Brown

SENIOR DESIGNER
Pete Ivey

PHOTO EDITOR
Jeffrey Braverman

VIDEOGRAPHER
Nat Wilson-Heckathorn

FABRICATOR
Daniel Spangler

WEBSITE

DIRECTOR OF WEB DEVELOPMENT
Parker Thomas

WEB DEVELOPER
Jake Spurlock
jspurlock@makezine.com

WEB DEVELOPER
Cole Geissinger

WEB PRODUCERS
Bill Olson
David Beauchamp

CONTRIBUTING ARTISTS
James Burke, James Delaney, Nate Van Dyke, Christina Empedocles, Damien Scogin

CONTRIBUTING DESIGNERS
James Burke

INTERNS
Kelley Benck (engr.), Eric Chu (engr.), Paloma Fautley (engr.), Sam Freeman (engr.), Andrew Katz (jr. engr.) Gunther Kirsch (photo), Raghid Mardini (engr.), Brian Melani (engr.), Nick Parks (engr.), Eloy Salinas (engr.)

SALES & ADVERTISING

SENIOR SALES MANAGER
Katie D. Kunde
katie@makezine.com

SALES MANAGER
Cecily Benzon
cbenzon@makezine.com

SALES MANAGER
Brigitte Kunde
brigitte@makezine.com

CLIENT SERVICES MANAGER
Miranda Mager

CLIENT SERVICES MANAGER
Mara Lincoln

EXECUTIVE ASSISTANT
Suzanne Huston

FINANCE CONTROLLER
Kevin Gushue

COMMERCE

VICE PRESIDENT OF COMMERCE
David Watta

DIRECTOR, RETAIL MARKETING & OPERATIONS
Heather Harmon Cochran
heatherh@makezine.com

MAKER SHED GRAPHIC DESIGNER
Uyen Cao

OPERATIONS MANAGER
Rob Bullington

PUBLISHED BY

MAKER MEDIA, INC.
Dale Dougherty, CEO

TECHNICAL ADVISORY BOARD
Kipp Bradford, Evil Mad Scientist Laboratories, Limor Fried, Joe Grand, Saul Griffith, Bunnie Huang, Tom Igoe, Steve Lodefink, Erica Sadun, Marc de Vinck

MARKETING

SENIOR DIRECTOR OF MARKETING
Vickie Welch
vwelch@makezine.com

MARKETING COORDINATOR
Meg Mason

MARKETING COORDINATOR
Karlee Vincent

MARKETING ASSISTANT
Courtney Lentz

MAKER FAIRE

PRODUCER
Louise Glasgow

MARKETING & PR
Bridgette Vanderlaan

PROGRAM DIRECTOR
Sabrina Merlo

BUSINESS DEVELOPMENT MANAGER
Heather Brundage

CHANNEL MANAGER
Kaitlyn Amundsen

PRODUCT DEVELOPMENT ENGINEER
Eric Weinhoffer

MAKER SHED EVANGELIST
Michael Castor

CUSTOMER SERVICE

CUSTOMER CARE TEAM LEADER
Daniel Randolph
cs@readerservices.makezine.com

Manage your account online, including change of address:
makezine.com/account
866-289-8847 toll-free in U.S. and Canada
818-487-2037,
5 a.m.–5 p.m., PST

Comments may be sent to:
editor@makezine.com

Visit us online:
makezine.com

Follow us on Twitter:
@make @makerfaire
@craft @makershed

On Google+: google.com/+make
On Facebook: makemagazine

MAKE SPECIAL ISSUE: Ultimate Guide to 3D Printing 2014 is a supplement to MAKE magazine. MAKE (ISSN 1556-2336) is published bi-monthly by Maker Media, Inc. in the months of February, April, June, August, October, and December. Maker Media is located at 1005 Gravenstein Hwy. North, Sebastopol, CA 95472, (707) 827-7000. SUBSCRIPTIONS: Send all subscription requests to MAKE, P.O. Box 17046, North Hollywood, CA 91615-9588 or subscribe online at makezine.com/offer or via phone at (866) 289-8847 (U.S. and Canada); all other countries call (818) 487-2037. Subscriptions are available for $34.95 for 1 year (6 issues) in the United States; in Canada: $39.95 USD; all other countries: $49.95 USD. Periodicals Postage Paid at Sebastopol, CA, and at additional mailing offices. POSTMASTER: Send address changes to MAKE, P.O. Box 17046, North Hollywood, CA 91615-9588. Canada Post Publications Mail Agreement Number 41129568. CANADA POSTMASTER: Send address changes to: Maker Media, PO Box 456, Niagara Falls, ON L2E 6V2

Go from Zero to Maker with David Lang!

A BRIEF HISTORY OF PERSONAL 3D PRINTING

WRITTEN BY **DALE DOUGHERTY**

Gregory Hayes

The first patent for 3D printing was obtained by Charles Hull in 1986, the same year that laser printers were patented and became available. Hull went on to found 3D Systems, which became one of the leading companies in industrial 3D printing. The essence of 3D printing is that software can "slice" a 3D image into a stack of 2D layers, and a machine builds the object by printing one layer on another. Sometimes called "additive manufacturing," 3D printing sounds like magic. It can be mesmerizing to watch a printer (by most definitions, a kind of robot), but it's also tediously slow, taking anywhere from 20 minutes to 20 hours to build objects that will fit in the palm of your hand.

After nearly two decades of industrial use, the personal 3D printing revolution started in 2005 with an open source project known as RepRap. Its overriding goal was to create a machine that could replicate itself. Its small but avid community of developers came up with a series of "evolution-themed" machine designs (Darwin, Mendel), while others created a software toolchain that could take .stl (for stereolithography, a term coined by Hull) files that describe a 3D model and generate the so-called "g-code" instructions that tell the machine what to do. The RepRap project, however, initially seemed to have trouble producing a machine that humans could build, let alone a self-replicating machine. (A maker who recently built a current RepRap model was able to print about 40% of the parts for his machine on another 3D printer.)

Frustrated by the process of building a 3D printer and realizing others were having the same problems, Bre Pettis, Zack "Hoeken" Smith, and Adam Mayer founded MakerBot in 2009. They sought to provide a kit version based on the RepRap designs but with enough improvements that they were easier to build.

The difference between RepRap and MakerBot was somewhat between baking from scratch and buying a box with all the measured ingredients you need to make something — and the first product from MakerBot was called the Cupcake. This kit took hours to assemble, and even then it wasn't perfect, but it was better than other options available at the time. The buyer of a Cupcake still needed special skills and patience to make it work properly. Other makers of 3D printers took the Cupcake as inspiration that they, too, could build a better printer.

In 2012, several changes began happening. MakerBot began working on a new version of a printer that would come fully assembled. And they began getting competitors from above and below. From above, manufacturers of industrial printers started producing their own personal printers, such as the Cube from 3D Systems. From below, others who were inspired by MakerBot's success began creating their own models, often using Kickstarter to raise money for development and production. Some of them were uninteresting, "me-too" rip-offs promising the same features and functions at a lower price. Yet more of the entrants offered solid competition, such as Ultimaker, MakerGear, Printrbot and Afinia. Still, even though there were more models and more manufacturers, demand exceeded supply.

In last year's *Ultimate Guide to 3D Printing*, we reviewed 16 different models of personal printers. This year, we tested 22 models, all of them coming to us assembled rather than as kits. Many of the personal printers do a good job at printing objects. While there is some differentiation in design and performance among the models, the biggest factors for usability are the software and documentation, which still favor the enthusiast willing to spend lots of time figuring things out. Buyer beware: The out-of-box experience is not at all like that of ordinary printers. The best way to have success in printing is to ask others in the community for help. The maker community has been key to the growth of 3D printing, providing not only tech support but also improving software and hardware.

With a price averaging around $2,000 and with capabilities that come close to their upscale competition, personal printers are in the hands of makers who are exploring the potential to turn rough ideas into real objects. Are you ready to join the 3D printing revolution? When you do, what will you make? ◢

Dale Dougherty is founder and CEO of Maker Media.

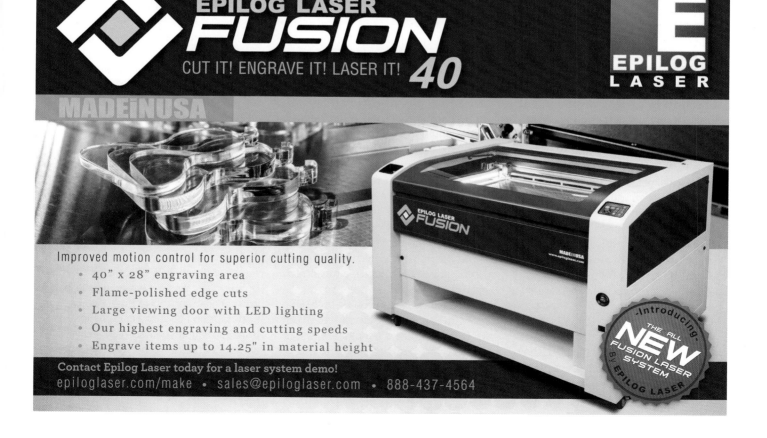

WHOSE PROBLEM IS IT?

If you make a tool and sell it to someone who goes on to break the law, should you be held responsible?

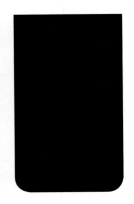

WRITTEN BY **CORY DOCTOROW**

Let's start by getting the question right. The wrong question is, "Will 3D printers be used to infringe copyright, trademark, and patents?" The answer to that is an emphatic and unequivocal yes. The right question is "Whose problem will this be?"

Whether you're making printers; distributing plans for printers; hosting a site for model files; printing things that people bring you; or even processing payments° for 3D prints, printers, or feedstock, you are a potential target for people who are upset about their copyrights, patents, and trademarks.

In 1976, Sony got sued for unleashing the first VCR upon the world. They'd advertised it as a way of recording feature-length movies from TV to tape for watching and taking over to friends' houses. The big movie studios, led by Universal, spent the next eight years fighting Sony over the device's existence. Now the studios didn't say that the VCR actually infringed on their copyright. A VCR is not a copy of a movie. Instead, they said that *some of the things that people could do with VCRs infringed their copyright*, and for that reason the VCR should be banned.

The courts went back and forth on this for years. If you make a tool and sell it to someone who goes on to break the law, should you be held responsible? It's not a standard to which we hold toolmakers usually, and that's a good thing, because cars, hammers, kitchen knives, and computers are all routinely used to break the law, and no one would be able to afford to make or sell them if they had to be responsible for what their customers did with them. In 1984, the Supreme Court narrowly ruled in Sony's favor and laid down the "Betamax rule": If a device is capable of sustaining a substantial noninfringing use, then it is lawful to make and sell that device. That is, if the device is merely capable of doing something legit, it's legal to make, no matter how it's used in practice.

But by the mid-2000s, the Betamax ruling was in tatters. A case against Napster took a huge bite out of Betamax. The judge established the principle that if you had knowledge of someone misusing your product and the ability to stop it, you are legally obliged to take action. This standard flies in the face of existing practices. It

means, in theory, that if you told Xerox that Bob's Copy Shop was letting kids use its leased photocopiers to make too many Garfield photocopies, they'd be legally obliged to call in Bob's lease and take away his copier. Thus far, this rule has only been used against peer-to-peer software companies, but as they say, bad cases make bad laws.

And worse cases make worse laws. In 2005, the Supreme Court ruled on the case of Grokster, another P2P company, whose technology had been carefully designed so that the operators of the software couldn't know what its users were doing, nor stop them from doing it if they found out. In Grokster, the Supremes cooked up a new theory of "inducement," holding that companies that encourage users to break the law were as guilty as the lawbreakers themselves. At the time, the studios claimed that this left the Betamax principle intact — Sony could have made the VCR and gotten away with it, even under this tighter version of the law. But that's just not true. Sony advertised its product for infringing purposes (recording movies and sharing them), and if Grokster's inducement standard had been the rule in 1984, it would've been the end of the VCR.

Meanwhile, in the world of user-generated content, the rule to watch has been "notice and takedown," which is set out in 1998's Digital Millennium Copyright Act. Under that rule, online service providers (from YouTube to Thingiverse to Dropbox to a message board to your college's student web pages) are not responsible for making sure that all the content that gets posted conforms to copyright law. But if you run a service where the public can post things, and you're notified that something infringes copyright, you're required to remove it immediately, or you can be sued alongside your user as an accomplice to the copyright infringement.

The notice-and-takedown policy is broken — most service providers treat any notice as gospel and immediately remove anything named in a notice, whether valid or not. That means it's easy to censor the internet through malice or incompetence (Fox Studios

You may just be a broke little maker in a garage, but if you've got deeper pockets than the user downstream of you who is breaking the law, be prepared for wealthy, powerful, entrenched industry to come after you.

filed fraudulent takedown notices on my novel *Homeland*, having confused it with their TV show of the same name). But it may get worse; Viacom's ongoing lawsuit against YouTube argues that the company should somehow proactively remove any copyright infringements from the 100+ hours of new video it receives *every minute*, and be held responsible for any infringing material a YouTube user puts online. Such a principle would kill virtually every successful internet platform today — there's no way Twitter, Facebook, Google, Dropbox, or even your college webserver could premoderate all the things their users want to put online before they go live.

What does this have to do with 3D printing? Everything. Actual infringers — users of printers and services — are hard to find, kind of clueless, and often broke. Companies have registered addresses, bank accounts, and offices where papers can be served. Users are numerous and diffuse. Companies are handily concentrated. Even more concentrated are investors, and even though the concept of a limited liability company should theoretically protect them from being named in a lawsuit, this didn't stop the record labels from suing the venture captalists who

invested in Napster, *and* the companies that gave those VCs their money. This didn't withstand legal challenge, but that wasn't the point — the point was to scare the pants off of a bunch of insurance- and pension-fund managers and get them to give orders to VCs to stay away from P2P, which they proceeded to do.

You may just be a broke little maker in a garage, but if you've got deeper pockets than the user downstream from you who is breaking the law, be prepared for wealthy, powerful, entrenched industry to come your way. As the jogger said to his slow friend: "I don't have to outrun the bear, I just have to outrun *you*."

What can you do about this? Well, I quit my P2P startup job and went to work at the Electronic Frontier Foundation. That may be outside of your immediate plans (they can't hire all of us!), but they're our best bet for sane laws and policies — for beating judges and politicians and regulators over the head with the clue stick until they understand that extending liability beyond users to services, companies, manufacturers, and other intermediaries is a dead end. ∎

Cory Doctorow (craphound.com) is a science fiction author, activist, journalist, blogger, co-editor of Boing Boing (boingboing.net), and the author of the bestselling Tor Teen/HarperCollins UK novel *Little Brother*. His latest young adult novel is *Homeland*, and his latest novel for adults is *Rapture of the Nerds*.

UP Plus 2

with Auto Calibration

3D printing simplified

Designed & Manufactured by Tiertime
www.PP3DP.com

10 COOL 3D PRINTED OBJECTS

The Photoz — Zung

"Open Wings" Cape

MELINDA LOOI
With the help of Belgian 3D printing company
Materialise, Malaysian fashion designer Melinda
Looi 3D printed her bird-inspired runway collection.
makezine.com/go/wingscape

Laser-Sintered Rocket Nozzles

NASA is testing the limits of 3D printing with
complex, nickel-chromium alloy rocket injectors.
makezine.com/go/rocketinjectors

hildendiaz

Light Sculpture Lamp

HILDEN & DIAZ
Artists are finding interesting ways to use 3D
printers, like Forms in Nature, a light sculpture that
casts fantastic shadows of gnarly trees on the walls
of the room in which it hangs. makezine.com/go/
lightsculpture

3D Print Yourself

There are services out there now that will 3D scan and print you as a figurine, like these high resolution captures by Twinkind (twinkind.com). Get started making basic 3D figurines with our 3D Scanning Party "Photo Booth" on page 44.

Nanomolecular Art

SHANE HOPE
Artist Shane Hope reshapes 3D models of nanoscale structures, like protein molecules and DNA double helices, by running Python scripts to visually "glitch out" the structures. He then prints the forms on clear acrylic and fills in empty spaces with paint. shanehope.info

2D Drawings into 3D Prints

Bring your child's whimsical drawing to 3-dimensional life by turning it into a 3D print, with the help of services like Crayon Creatures (crayoncreatures.com).

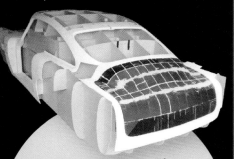

Aston Martin Replica

IVAN SENTCH
Rather than shaping by hand, New Zealander Ivan Sentch is 3D printing the mold plug for the body of his replica 1961 Aston Martin, 4" by 4" at a time.
makezine.com/go/astonmartin

Dodecahedron Speaker

SEAN MICHAEL RAGAN
This array of 12 speakers, arranged on the faces of a 3D-printed dodecahedron, provides non-directional-speaker quality sound. makezine.com/go/dodecahedron

3D Artwork for the Blind

The Midas Touch project transforms well-known two-dimensional artwork into 3D, allowing blind people to "see" it with their fingers.
makezine.com/go/midastouch

TARDIS Prime Transforming Toy

ANDREW LINDSEY
Inspired by a fanciful T-shirt depicting a Doctor Who TARDIS transforming into an Optimus Prime-style robot, New Jersey engineer Andrew Lindsey designed and printed this amazing modular TARDIS Prime. thingiverse.com/thing:106595

Gunther Kirsch

JUST WHAT I NEEDED

3D prints to the rescue!

"To my surprise I found that the bridge on my electric ukulele ripped off. I could get it repaired or the whole uke replaced, which would have left me for weeks without my uke. Measuring the original dimensions, minutes of modeling and a half an hour of printing was all it took to have a prototype. Four design iterations later, the bridge was beefed up and the cord fixation changed just to my liking. I have now a better uke bridge than before!"
— *Martin A. Koch, Manresa, Spain*

Greg Williams

"My office has a paper towel dispenser, but the key was long gone. I designed and printed a new key. Since I couldn't find a replacement key, it saved me from having to replace the entire dispenser."
—*Greg Williams, Milltown, N.J.*

"I made a small wire spool rack with an old broom handle, but my attempt at organization soon became a lesson in knot theory — an utterly tangled mess. Rubber bands and tape were too tedious and sticky, so I worked up a little parametric clamp in OpenSCAD to help keep spooled wire in its place. Just pull it off to get to the wire, and snap it back in place when you're finished. Or you can pull the wire through the window while holding the tab."
—*Alex Franke, Chapel Hill, N.C.*

"My boy scored a nice soccer goal, and when I sat back down, I went falling through the back of my folding travel chair. I can't imagine anyone would be able to sell me the part to fix it, so I turned to OpenSCAD and made my own. It's held up very well for over a year now, and soon I'll be printing another one to repair a similar tear on the other side."
—*Alex Franke, Chapel Hill, N.C.*

"I'm with SYN Shop, the Las Vegas hackerspace, and we just started doing a podcast. I bought an inexpensive Canon video camera (as an extra) and wanted to use my Sony wide-angle lens on it. Needless to say, the lens adapter that came with it didn't fit. So, I printed one. The cool thing is that it's held onto the camera with friction only, since the Canon doesn't have a threaded lens ring. I got to learn a little about Blender, and this was also my very first 3D print!"
—*Bill Tomiyasu, Las Vegas, Nev.*

"The door of my refrigerator has some

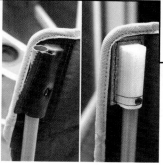

Alex Franke

Bill Tomiyasu

rails on the inside shelf area that are supposed to keep the mustard from flying when you open the fridge. One of mine broke, and naturally, the little hook bit fell into the door itself, never to be recovered. So I took the mirror image one from the other end of the rail, modeled up the part I needed (tricky little geometry) in SolidWorks, and printed it in ABS. Problem solved! You'd be hard-pressed to even notice that the part isn't original."

—*Eiki Martinson, Lighthouse Point, Fla.*

Eiki Martinson

"When we moved into our current house, the light above the front door was just a bare bulb, with nothing around it. While at Home Depot, the wife found a cheapie plastic thingamajigger that would fit over the light bulb. This is also when she suggested I could 'make it work.' I really don't know how I could have done this as elegantly without a 3D printer. Using open source software I designed the needed adapter and then printed it out using open source hardware, and the total cost of materials (ABS plastic) was probably less than 50 cents." —*Pete Prodoehl, Milwaukee, Wis.*

Pete Prodoehl

"I do some video work for my band and other small nonprofessional side jobs. I've always wanted a rail system for my DSLR but just couldn't justify the expense of some of this stuff for a hobby — a camera rail system with a follow focus can cost upwards of $1,000 or more! So I decided to build my own using my 3D printer and parts I could find at the big box stores. I designed some parts, used Marcus Wolschon's Follow Focus off of Thingiverse, and got exactly what I needed for less than $30!"
—*Jamie Cunningham, Clearwater, Fla.*

Jamie Cunningham

"The BuddyGripper3D evolved from the BuddyGripper Original (buddygripper.com), which was developed while I lived in Madrid, Spain. There on a Fulbright Grant to finish my Ph.D. in mechanical engineering, I realized I needed a device to attach my iPhone to a tripod so that I could travel and capture my experiences. Manufacturing was costly, so I switched to 3D printing. After that, creativity flourished and I shared it with the Thingiverse community. What better way to 'put fun in your pocket' than to 3D print your own Buddy-Gripper3D?" —*Chris McCoy, San Francisco, Calif.*

Chris McCoy

LIFE-CHANGING 3DPRINTS

WRITTEN BY
CRAIG COUDEN

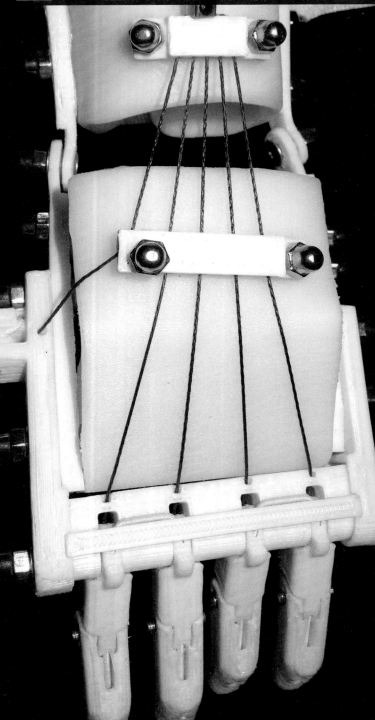

Robohand

After cutting off four fingers in a woodworking accident, Richard Van As saw a need for affordable mechanical prosthetic hands. Using two donated MakerBots, he and 3D designer Ivan Owen came up with Robohand, a custom appendage for kids missing fingers. Best of all, the designs are open source and available on Thingiverse — when outgrown, simply reprint. robohand.net

Kaiba's Tracheal Splint

Doctors from the University of Michigan 3D printed a tracheal splint to stabilize baby Kaiba's underdeveloped air passageway. The bioplastic implant will be absorbed into the body as Kaiba's airways strengthen. makezine.com/go/trachealsplint

Scaffolds to Grow Joint Cartilage

Researchers from the University of Wollongong and St. Vincent's Hospital Melbourne are developing 3D-printed scaffolding to help regrow cartilage to treat osteoarthritis and traumatic injuries. makezine.com/go/cartilage

Organ Models

Jewelry and design company M.C. Ginsberg partnered with doctors at the University of Iowa to print models of patient organs before surgeries. Based on CT scans, surgeons can use the anatomically accurate models to plan parts of surgery before walking into the operating room. makezine.com/go/3dprintedorgans

Bones

From metal vertebrae and cranial plates to replacement jawbones and even teeth, 3D-printed bone replacements are in the process of becoming bespoke alternatives to other biomedical implants.
makezine.com/go/skullplate

Oxford Performance Materials' OsteoFab implants are shaped to each patient's anatomy.

Skin Cells for Burn Victims

Researchers at Wake Forest University hacked an inkjet printer to lay down living cells on top of burn wounds. After scanning the wound, the experimental printer deposits two different kinds of cells from hacked ink cartridges, which form the dermis and epidermis layers of new skin.
makezine.com/go/3dprintedskin

Bionic Ear

Princeton researchers used live cells, silicone, and silver nanoparticles to develop a proof of concept bionic ear that can detect frequencies beyond human hearing — even radio waves!
princeton.edu/~mcm

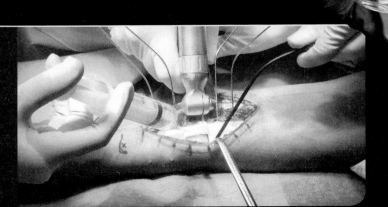

Surgical Guides

Doctors used Materialise's Mimics Innovation software to 3D print surgical guides to more precisely operate on a man's broken arm. Printed from a CT scan, the guides directed cutting and drilling points and reduced time in the operating room.
makezine.com/go/surgicalguide

GROWING UP & GOING PRO

WRITTEN BY **BOB PARKS** *and* *ILLUSTRATED BY* **CHRISTINA EMPEDOCLES**

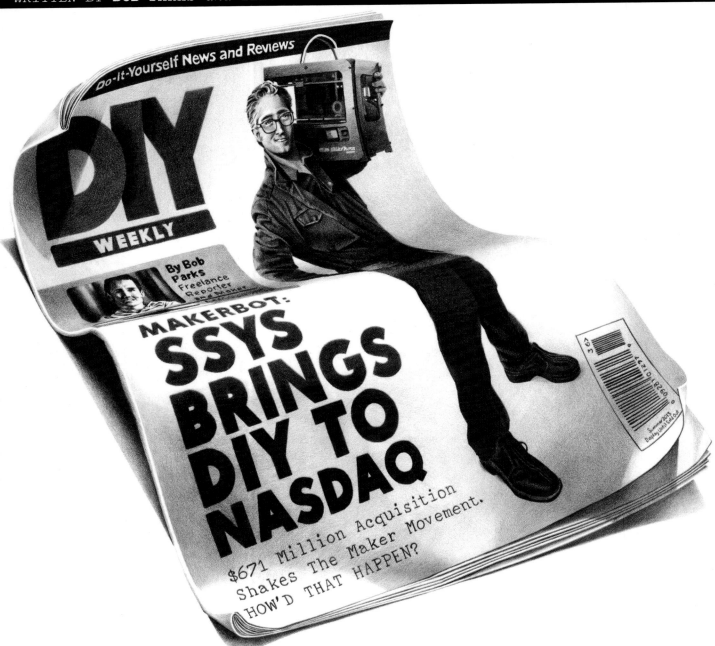

Do-it-Yourself News and Reviews

DIY WEEKLY

By Bob Parks
Freelance Reporter and Maker

MAKERBOT: SSYS BRINGS DIY TO NASDAQ

$671 Million Acquisition Shakes The Maker Movement. HOW'D THAT HAPPEN?

Paul Bohm, founder of hackerspaces.org, was reading a news feed on the subway when he heard the MakerBot news. "I was immediately excited for them," says the 30-year-old entrepreneur. "Not because someone somewhere made money, but because it's a case study in taking an idea, hiring people in your own local area, and executing on it."

When word that MakerBot — the Brooklyn-based 3D printer manufacturer — was acquired on June 19, many garage entrepreneurs like Bohm had been following the story real-time. It started out as a project among friends in the NYC Resistor makerspace, a simple effort to build a flexible, easy-to-use version of the open-source RepRap printer. An early test unit in 2008 sent up smoke, but after many tries and five generations of printers, the company now has 330 employees, sales of over 10,000 U.S.-made bots, and projected revenue of more than $50 million.

With growth brought many changes. Along the way, the company that started with an adamantly open-source philosophy found that opening all their technology to the public wasn't a viable business, a move that offended many in its user community. Then amid conflicts, two of the three company founders left

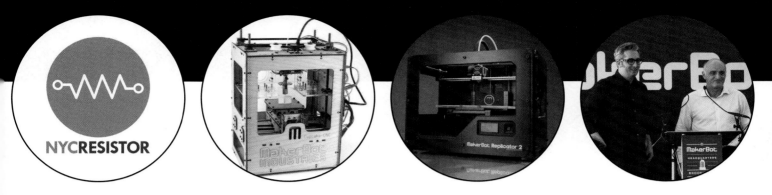

NYCRESISTOR

in 2012. And as MakerBot rose and early investors sought a return, it became an attractive acquisition target for a large industrial firm that — 25 years ago — pioneered this kind of 3D printing, Minneapolis-based Stratasys.

The acquisition, 100% of MakerBot in exchange for 6.9 million shares of Stratasys (3.9 million offered right away and 3.0 million based on MakerBot's performance between now and the end of 2014), brought opportunities to reach more users as well as financial rewards to the 'botters. Securities and Exchange Commission documents list co-founder Bre Pettis, 41, as earning around $151 million in stock. His former partners, Zach "Hoeken" Smith, 29, and Adam Mayer, 37, will earn around $103 and $94 million respectively. Adrian Bowyer, 61, who created the RepRap project and invested early in the company, is slated for a $7 million payout.

How did a hacking project become a coveted Wall Street property? MakerBot was the first easy-to-use, low-cost 3D printer — and although this terrain is usually connected with consumer users, the concept had a whole different ring to Stratasys, which began to observe corporate engineers and garage inventors buying desktop units to print out rough drafts of designs before they went to more expensive 3D printers.

For instance, Ford engineers in Dearborn, Mich., have a handful of Replicator 2s they use to rough-out car parts. "It's a great tool to get a looks-like, feels-like model," says Zach Nelson, a Ford engineer who has had three generations of MakerBots on his desk at various times. Stratasys CEO David Reis, in a recent conference call with market analysts, said, "We believe this trend is similar to the evolution of personal computers.

What started as kit-based products for enthusiasts became a mainstream tool in business and industry as affordability, access, and ease of use improved."

With $359 million in revenue in 2012, the Minneapolis company's desire to release an affordable printer may have also been prompted by its chief rival, 3D Systems of Rock Hill, S.C. (itself with $354 million in revenue). "Additive manufacturing has finally reached an inflection point," says Pete Basiliere, a research director for Gartner. "The early innovators are trying to pull together a

> ## How did a hacking project become a coveted Wall Street property?

complete ecosystem to make theirs the standard approach." For starters, both industrial companies have recently purchased or built online repositories for 3D files that help users get started on designs (MakerBot came with high-traffic Thingiverse.com, a free site of more than 100,000 user-generated objects).

The two companies are also pushing lower-cost hardware. Previously, Stratasys' cheapest printer was $9,900 (3D's was $10,990), but then 3D bought the RepRap-based 3D printer company Bits from Bytes in 2010, and then launched the $1,300 Cube in 2012. Just as MakerBot has industrial users of its desktop printers, 3D Systems has sold Cubes to engineers at GE Aviation to explore the potential of quick-and-dirty desktop prototyping for aircraft parts.

Stratasys has said it intends to keep MakerBot an independent subsidiary that will benefit from the advanced materials science and R&D of its

professional-grade machines. Currently, however, gross margins on MakerBot's products are lower than that on the corporate devices. One issue may be that Replicator 2s currently run on either MakerBot's own low-cost plastic filament or any common supplies. (In contrast, the consumer devices from 3D Systems use a chip-embedded cartridge so users can only buy the higher-cost certified filament.) It is impossible to predict how Stratasys will handle this issue. In an earnings call, CEO Reis said, "We're working on long-term plans for the business model of MakerBot to create a more similar business model to the current one of Stratasys."

One thing is clear, in running the MakerBot subsidiary: Pettis and other managers now have to make decisions with an eye to the needs of the whole company and its shareholders. It's a classic startup dilemma — if you seek early investor funding, you will soon either become a public company or likely become acquired by one.

"The transformation has been fascinating to watch as a 'maker company' that started a little before MakerBot," says Limor Fried of Adafruit, an online electronics seller with $20 million in annual revenue. "We've thought about funding a few times, and what it would mean if we had gone that way. Once you take on investment, you end up making different decisions, hiring specific people for an exit/sale, and buying into a system that's very different than going at it on your own. You have more freedom the longer you can hold out, but, then again, it's a lot of freedom to go out of business." ∎

Bob Parks (judder.net) is a freelance reporter and a maker.

Mark De Vinck (cupcake), MakerBot® (Replicator 2, Pettis and Reis)

INSIDE THE NEW MAKERBOT

Bre Pettis talks shop. INTERVIEW BY MARK FRAUENFELDER

Photo courtesy of MakerBot

Bre Pettis, a former schoolteacher (and producer of MAKE's popular "Weekend Projects" video series), recently sold his 3D printer company MakerBot to Stratasys. We talked to Pettis to find out what he's working on, why he sold his company, and what his plans are for the future of 3D printing.

Tell me about your most recent product, the Digitizer, a 3D scanner that sells for $1,400.
I really wanted to have a 3D scanner — something that would be easy, that would get the job done, and that would be as affordable as possible. What was out there started at about $3,000 and went up. It turns out that the hardware on this is not terribly hard. You have a turntable that moves around. You have two lasers. We use nice line lasers — quality does matter on this. And then you have a webcam and a really fancy filter, which filters for 650nm. The idea is that the camera will only see the color of the laser line.

It's just a beautiful machine. I'm really proud of it. We finished it in July 2012, and it took us that long to develop it from a project into a product. When you're dealing with big injection-molded pieces, the tooling costs more than $100,000. You can only really justify doing that when you know you're going to sell tens of thousands of them. If you're not going to do that, you might as well stick with laser-cut wooden parts.

One of the cool things about this is it's also made for manufacturing. The screws on this are all on the top so that it never needs to be flipped over. There are little tricks like that that mean that we can just crank these things out.

Does it offer the same resolution as the Replicator 2 — 100 microns?
Because a scanner turns an object and the laser fires at it, the resolution is radial. It actually does 800 different points around a circle. While the layer resolution is important on a MakerBot Replicator 2, the radial resolution is what's important on the Digitizer. It ends up with a model that's usually more than 200,000 polygons, which is pretty darn good.

In June you were acquired by Stratasys, which was a really nice sale. You now have a great amount of stock in Stratasys. It's a testament to the company you built and the great products you've developed. Could you tell me the reason why you sold it to Stratasys, and how that's going to help you outpace the competition?
We started MakerBot before Kickstarter, so we had to give up equity in the company to get our original $75K investment. And then, as we grew, we needed more equity so we could continue to develop the machines. I raised $1.2 million from angel investors. Then in 2011, I raised $10 million from the best venture capitalists. But when you do that, you're making a promise. You're saying, "You're going to give me money so I can grow the company, and in return at some point, what you've invested is going to be worth more than you put in."

I had been on track to go around and raise another round, but in the middle of it, Stratasys expressed an interest, and because they're sort of the grandfathers in the space, they were really the only people we would have considered merging with. They're

also just good people. We got into the thick of it, negotiated it, made it work. It's interesting, we get resources now. Rather than me going out and raising money, I have the resources of a public company to dip into.

One of the things that I don't know if many people realize is that we spent a lot of time at MakerBot routing around the intellectual property of the big companies. There are probably around eight patents that we couldn't have that we had to work really hard to route around to be able to have our products. Now we don't have to stress out as much about it. We actually get access to IP that we didn't have before.

Is that because Stratasys holds those patents?
Stratasys has about 800 patents in the space.

There are some other 3D patents, such as laser sintering, which are due to expire soon. Are you looking into how you might incorporate them into what you're doing?
Yes, it's interesting. From the beginning, we could only do what we did because the original patent expired. But since then, there have been a lot more patents. It's one of those things where we had to decide early on that we were going to be a sustainable company, that we were going to grow, and we were eventually going to be a big company. So we had to invest in intellectual property because the patent system is really weird — you want to basically have enough IP that you can survive.

We have a patent on our automated build platform. We have some networking patents. We have some interesting stuff that basically allowed us to get ready if we ever had to go head-to-head with anybody. But the acquisition sort of solved that as well. It turned out really well for everybody. ◪

Watch the video of our complete interview with Bre Pettis at makezine.com/go/bre-interview.

PATENT WATCH

Want to launch a groundbreaking desktop 3D printer? Better lawyer up.

Over the last five years we've seen an explosion of inventive 3D printers designed and built by makers in hackerspaces, but the next generation of these machines will likely be guided by courtroom decisions. As the 3D printing market becomes more lucrative, intrepid entrepreneurs will need to navigate a minefield littered with thousands of patents that are owned by entrenched competitors, any one of which could lead to an epic print fail. Here are four such patents, giving a preview of a couple technologies that are soon becoming available to makers, and two new ones that may be locked up for decades to come.

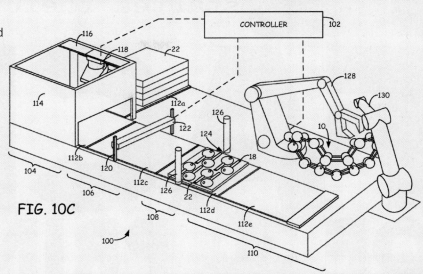

FIG. 10C

EXPIRING PATENTS

Method and Apparatus for Prototyping a Three-Dimensional Object
- **Patent Number:** US6007318 A
- **Date Issued:** Dec. 28, 1999
- **Assignee:** ZCorp
- **What It Means:** This patent covers the production of 3D models by depositing ink and glue on a plaster powder base. Its entry into the public domain means that Minecraft fans will soon be able to transform their blocky dream worlds into vivid, full-color — if slightly dusty — models, provided some chalk-loving creator gets busy developing a sandy solution.
- **Who Could Stop It:** Hewlett-Packard, the king of 2D printing. This type of 3D printer requires robust ink jets to deposit glue and ink. HP has a treasure trove of patents covering this exact technology and isn't likely to share them with just anyone.

Method and Apparatus for Production of Three-Dimensional Objects by Stereolithography
- **Patent Number:** US5554336 A
- **Date Filed:** March 11, 1986
- **Assignee:** 3D Systems
- **What It Means:** Stereolithography is essentially high-def 3D printing. Using laser and UV-sensitive resin, printers using this technology can produce parts with resolution four times greater than those from machines like MakerBot's.
- **Who Could Stop It:** 3D Systems isn't shy about suing companies it feels are infringing on the thousands of patents that protect their inventions, and the $5.4 billion dollar company has an army of lawyers that can hunt after companies they think are potentially infringing on their patents. Just ask Formlabs.

RECENTLY FILED PATENTS

Combined Process for Building Three-Dimensional Models
- **Patent Number:** US20100161105 A1
- **Date Filed:** Dec. 18, 2009
- **Assignee:** Stratasys
- **What It Means:** The next generation of 3D printers could fuse additive and subtractive processes — basically they can deposit material during one phase, then cut it away in a secondary process. This opens the possibility for more complex shapes, a wider variety of materials, and new applications.
- **Who It Affects:** The error prone. This patent is essentially "Control+Z" for plastic. Imagine if someone patented the "delete" key on your keyboard!

Seam Concealment for Three-Dimensional Models
- **Patent Number:** US8349239 B2
- **Date Issued:** Jan. 8, 2013
- **Assignee:** Stratasys
- **What It Means:** This patent doesn't cover the machine, but rather the way the machine deposits plastic. 3D printing software that utilizes the algorithm disclosed in this document can hide the unsightly seam that occurs in the fused filament fabrication process. Machines that don't will be forced to create parts with ghastly gashes running down their lengths.
- **Who It Affects:** Anyone who wants to make models with smooth surfaces. This patent means hobbyist 3D printers will need to live with "scarred" models until 2030 at the earliest.

Meet seven makers who started their own companies.

WRITTEN BY **GOLI MOHAMMADI** AND **MIKE SENESE**

FACES
OF 3D PRINTING

DIEGO PORQUERAS OF BUKOBOT
deezmaker.com/bukobot

Necessity was the mother of invention for Los Angeles-based maker Diego Porqueras. While working as a Hollywood digital imaging technician, he was drawn to 3D printing for its potential to make custom camera brackets and adapters. He started attending the "MakerBot Monthly" meetups at Crash Space, his local hackerspace. After building his first printer kit, a Prusa Mendel RepRap from MakerGear, he confesses, "My perfectionism got in the way and I ended up creating my own, better 3D printer." He launched Bukobot on Kickstarter in April of 2012, hitting 400% of his funding goal, and quit his job to focus on making printers. By July, he added notable Rich ("whosawhatsis") Cameron, the creator of the Wallace RepRap printer, to his team.

Riding the momentum, in September of 2012 Porqueras opened Deezmaker, the first physical 3D printer shop and hackerspace on the West Coast, fueled by the desire to share the joy of actually seeing 3D printers in action. Porqueras has built his business on the belief that quality high-end machines can be made without the high-end price.

ANDREW RUTTER OF TYPE A MACHINES
typeamachines.com

Not only does Andrew Rutter care deeply about building machines that consistently deliver great results, he's equally as concerned with making printers that are easy to fix when they do fail. He explains, "We don't just want people to like the results, we want them to appreciate the design and engineering of the machine, and for that to be as important to them as the output." One of the leading minds in 3D printing, it's hard to imagine that previously, San Francisco-based Rutter spent roughly 10 years in the entertainment industry, "mostly as a lighting technician, occasionally as a designer, and once or twice even as a performer." He built his first printer, the ubiquitous MakerBot Cupcake, in December of 2009.

Rutter founded Type A with the desire to make a better machine, and in January 2012, the core team was born out of the Noisebridge hackerspace with the addition of Espen Sivertsen, Miloh Alexander, and Gabriel Bentley. Type A's team is now 15 people who proudly manufacture, assemble, and test their machines in their California workshop. Rutter has been responsible for getting 3D printers in his local public library and says of the future, "Whatever happens, I want to see a 3D printer on every student's desk."

MAXIM LOBOVSKY OF FORMLABS
formlabs.com

Five years ago, while studying applied physics at Cornell University, Maxim Lobovsky began working on the massive, transcontinental Fab@Home open-source, personal-fabrication project, which sparked his interest in 3D printing. After Cornell, he went on to become a researcher at the MIT Media Lab. In 2011, upon earning his master's at age 23, he and two fellow MIT grads, Natan Linder and David Cranor, founded the Cambridge, Mass.-based company Formlabs, makers of stereolithography resin-based printers. Lobovsky says, "I realized that there was nothing standing in my way — that I could bring all my ideas for a desktop 3D printer to a finished product. It didn't need to be a massive endeavor that only a huge company could take on, but something that a few dedicated people could bring to life."

After raising nearly $3 million in 30 days toward their original $100,000 goal, Lobovsky and the team knew they were onto something. He explains, "My passion is making the technology accessible to a wider range of users. There is a lot of amazing stuff possible beyond the FDM machines that people are becoming familiar with. For me, accessibility means both price point and ease of use."

BROOK DRUMM OF PRINTRBOT

printrbot.com

Former youth pastor and coffeehouse owner Brook Drumm discovered his 3D printing fascination in early 2011 (see "Generation 3D," page 30). A tireless man with a magnetic personality, Drumm set up a 3D printer meetup group in his area and helped foster the blossoming community. Quickly growing adept with existing technology through the group, he soon realized creating a low-cost machine was a realistic possibility.

"It hit me one night, very late, that ripping out all sorts of unsightly support could result in a sufficiently rigid structure that was a little more straightforward," he says. "After enduring yet another lengthy recap of my dreams to design and sell 3D printers, my wife, Margie, gave me the best advice to date — 'It has to be cute.'

"I dove into a complete and total design obsession for 6 months," he continues. "The Printrbot was a product of a young and eager community rapidly prototyping on RepRaps. A wild ride on Kickstarter sealed my fate."

With over $830,000 in pledges, Printrbot launched into the forefront of 3D printing in December of 2011. Since then, Drumm continues to embrace openness with his company, actively engaging with customers and even freely discussing Printrbot's future plans. When asked what he's working on, he rattles a laundry list of projects, including Kinect-based body scanners, DLP resin 3D printers, a desktop CNC with iPad CAD/CAM software, and designing and printing arms for needy kids in the Sudan. "It's gonna be a great year!"

RICK POLLACK OF MAKERGEAR

makergear.com

When Rick Pollack had an idea for a new product in 2003, he didn't know how to go about manufacturing it. Having worked as a software developer since the mid-90s, he explains, "I ended up spending a lot of time and money figuring out how to get it made. That planted the seed of desktop manufacturing — I wanted to be able to take an idea and hold it in my hand in hours."

In 2009, with the availability of low-cost electronics and open source tools, the time was right to launch MakerGear. Originally, he identified a need for better extruders, bought a small 70s lathe for $250, and started making extruder parts, one at a time, in his unheated garage in northeast Ohio. Today, his parts and printers are owned by 3D printer enthusiasts in 75 countries.

Pollack takes pride in the fact that MakerGear manufactures in the United States, with a vast majority of their custom components made by fabricators and machinists in Ohio. Though their off-the-shelf parts come from overseas, Pollack says, "We're working to bring even more of that manufacturing to the community. We've worked very hard to provide a quality machine at a reasonable price. We want people to be happy with the machine and enjoy working with us."

STEVE WYGANT OF SEEMECNC

shop.seemecnc.com

Indiana-based mechanical engineer Steve Wygant had been running his own machine shop since 1996, and for three years, he machined parts for large orthopedic companies, during which he first learned about 3D printing. He was fascinated by the stereolithography machines at one of the companies but knew they were out of reach for hobbyists. Then, in 2009, he learned about the RepRap project and was intrigued by the design and mechanical structures of the Darwin and Prusa machines. In 2011, Wygant (aka "PartDaddy") joined forces with John Olafson (aka "Oly"), an experienced CNC woodworker and machinist, combining their large and diverse skill sets on the first SeeMeCNC printer, a derivative of the RepRap Huxley, calling it the H1.

When asked about his latest "aha" moment, Wygant recalls working on their Rostock Max printer and accidentally inventing the "Cheapskate" linear bearing: "While holding a 608 skateboard bearing and T-slot aluminum, I accidentally fit the bearing to the side slot and realized I discovered a low-cost linear motion carriage."

Wygant and Olafson are passionate about empowering makers to bring their ideas to life and nurturing the vibrant 3D printing community, reveling in how many folks from different countries and backgrounds 3D printing unites. Wygant adds, "Seeing them all interact and work together to create amazing things is just awesome." They're proud sponsors of the Midwest RepRap Festival and support the new independent file-sharing site repables.com.

ERIK DE BRUIJN OF ULTIMAKER

ultimaker.com

Ultimaker's Erik de Bruijn can be called an old-timer in desktop 3D printing. Involved with the RepRap project since early 2008, he was one of the first in the community to successfully replicate functional parts, quickly followed by full machines. De Bruijn's background with RepRap, and with the IT company he founded at age 16 (still operational 12 years later), provided invaluable when developing his own 3D printer with partners Martijn Elserman and Siert Wijnia in 2011. "When we launched Ultimaker, my business experience was put to the test," he says. "Ultimaker turned out even more successful than we imagined, to a large part thanks to the great open-source community."

The Dutch-based de Bruijn has also helped establish key elements in 3D printing, including combining ABS and PLA, and helping develop retraction — when excess filament pulls back into the extruder to avoid stringing. He credits a meeting with another distinguished 3D printing developer as a part of this. "When I had a RepRap speaking engagement in the UK, I had the opportunity to visit the famous Nophead," he says. "He did most of the coding, I mostly encouraged and provided ideas and suggestions, but together we showed the community how much retraction matters. After that, it became a general feature."

With a brand new machine and a new design-sharing site called YouMagine, de Bruijn and the Ultimaker team are keeping busy, but he remains focused on keeping their creations open. "I believe people should be able to make their tools their own. While I don't know where other companies will take things, this is where Ultimaker is headed."

THE BUSINESS OF PRINTING

From generating models of ancient Greek and Roman sculptures to creating customized dolls, these makers share insight into what inspired them to get serious about printing.

3D-printed objects create opportunities for these four makers.

ALICE TAYLOR OF MAKIELAB

makie.me

"I had a eureka moment at the New York Toy Fair in 2010: The Digital Kids Conference was co-located with the Toy Fair, but the digital and the physical were far separated by long corridors and concrete walls. In the digital basement were the folks building huge digital worlds, full of characters, monsters, penguins — and upstairs was the cavernous hall full of characters, monsters, and toy animals. I had that moment of, 'Why can't we build dolls straight from avatars, using 3D printers?' Fast-forward a bit, and we have Makies, our build-a-doll series, which also happen to be the world's first (actually toy-safety-certified) 3D-printed toy."

Printer: "At work, we have two Replicators, and one Thing-O-Matic. We do our prototyping on the Replicators, but our dolls are printed in nylon powder (SLS) on an EOS series machine with our manufacturing partners."

✚ Taylor's stylish Makie Dolls are individualized by customers online, then printed and sent directly to the buyer.

Jeffrey Braverman

JOHN ALLWINE OF FREAKIN' SWEET KNOTS
knots.freakinsweetapps.com

"My dad's been teaching me how to tie knots since I was a kid. I've been able to tie a bowline backwards or forwards in seconds for as long as I can remember. I've been programming since I was a kid, as well, which I also learned how to do from my dad. I have a degree in computer science from Cal Poly, where I took as many 3D graphics classes as I could. Inspired by wanting a Turk's head knot engagement ring for my wife, I started by writing programs that generate instructions for how to tie the knots, which evolved into programs that generate 3D models of Turk's heads that can be 3D printed. Right now my Freakin' Sweet Knots app generates customized woven rings which can be ordered through Shapeways."

▣ **Printer:** "Not yet! I recently won an Instructables contest which earned me a MakerBot Replicator 2."

COSMO WENMAN, INDEPENDENT ARTIST
cosmowenman.com

"Last summer I tried drumming up public interest in my *Through a Scanner, Skulpturhalle* project to 3D-capture and freely publish 3D-printable models of ancient Greek and Roman sculptures. It didn't quite catch fire like I'd hoped, but luckily I found a corporate sponsor for the project in Autodesk. If all goes as planned, I'll soon have more concrete examples of what I was shooting for — like the first-ever publicly available 3D surveys of the *Venus de Milo*, *Winged Victory*, and the *Medusa Rondanini*, among others. I'm also supporting my scanning and publishing work by selling the first-ever 3D-printed iterations of the sculptures I capture and share. I sell the occasional large-scale, hand-assembled and finished 3D prints of those, and I also offer smaller prints."

▣ **Printer:** "A MakerBot Replicator 1 that refuses to die despite my not giving it even the most basic maintenance or adjustments."

KACIE HULTGREN OF PRETTY SMALL THINGS
prettysmallthings.com

"I work as a scenic designer in the theater industry. One part of my job is building scale models of set designs. I originally bought a 3D printer to incorporate in my model-building workflow. I soon realized that the designs I created filled a niche market, and my business was born. I sell scale model furniture, mostly 1:48 and 1:24 scale, to other designers and hobbyists. I started my business using my Replicator 1 to produce the inventory I sell in my web store, which helped me get started with very little overhead. Now I carry an inventory of prints, which has allowed me to expand and keep my desktop machines free for development."

▣ **Printer:** "My Replicator 1 and 2 are workhorses in my studio."

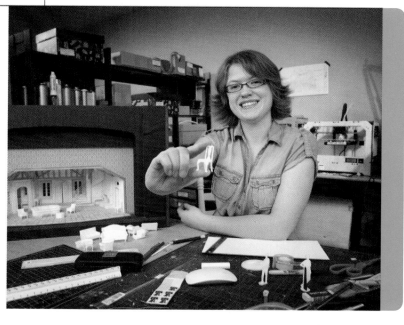

GENERATION 3D

Margie Drumm

WRITTEN
BY
**BROOK
DRUMM**

When I was a kid, I used to follow my dad around asking questions: "Why?" "What is that?" "How?" He had an office in the back of the house full of wonderful gadgets, and encouraged my desire to take everything apart to see how it ticked. He was always ready to explain what he knew about electronics and machines, which to a young boy sounded like pure magic. Being allowed to ask, explore, tinker, hack, and even fail set me on the course to being a maker.

I've carried on the tradition of encouraging my kids to ask questions and explore the healthy stash of electronics and gadgets I collect. Each of my three kids has tackled projects with me one-on-one, but we all converged on our first 3D printer in January 2011. I'd been inspired by the MAKE issue that featured Bre Pettis [of MakerBot, Volume 21] and saved up for months to buy my Cupcake 3D printer kit.

My daughters, Sydney and McKenna (at the time 13 and 11), helped me assemble various parts in the kit, and I taught my then-6-year-old son, Levi, to solder when it was time to finish the electronics. It was a rite of passage for him and me! He was both a little scared and excited to be allowed (under close supervision) to wield such a tool. It was obvious to me from the beginning that I needed to design an easier-to-build printer, but we eventually got the Cupcake up and printing. The whole family gathered around the kitchen table to watch that mysterious device crank out its first plastic blobs. Ever since, I've believed that 3D printing is a family sport.

That was the beginning of a new way of thinking in the Drumm household. My wife and kids began to see what was possible. They would see something at Target and say, "We can print this!" I started to fix things around the house — kitchen shelves, the washing machine knob, the toilet handle. My two daughters and I printed rings, bracelets, and doodads to hang on their key chains. I taught them to use SketchUp, a free 3D modeling program, and let them print what they designed.

Levi, now 9 years old, is starting to request more and more printed things.

3D printing opens a world of possibilities for kids.

In fact, his recent book report incorporated a printed pyramid, sphinx, and replica of the Rosetta Stone! I don't think he fully realizes the rare air he breathes. I wish I were there when he explained to his class what a 3D printer is and how it works.

It's normal for him to see printed items on the counter when he gets up in the morning, and I don't think he gives it a second thought. He regularly pitches his ideas of cars, robots, automated machines, and toy guns that we should design and print. I love that his ideas truly have no bounds. He doesn't fully grasp the laws of physics, but I think that's all the better. I absolutely love that he's grown up believing that you can print and build anything.

The other day he was watching me draw up a new printer design in SketchUp. He had been watching intently for some time and I saw his wheels turning, so I asked him some key abstract ques-

> "My son is growing up in a different world than I did with my 2D scrollers and text-based adventure games on the Commodore 64."

tions about the drawing. He rattled off all the right answers immediately. I was shocked. I never formally taught him any of these concepts, but he got it. It dawned on me that the video games he plays are in 3D space. He's growing up in a different world than I did with my 2D scrollers and text-based adventure games on the Commodore 64.

One of the things I love most about working with kids is that they don't know the limits of what is possible or what they can do. They have fresh eyes on problems, and if they're curious and passionate, they also have the time to solve problems. Caleb Cotter is such a person. Caleb first showed up to our RepRap meetup group two years ago.

He's 16 now, but was 14 at the time. His grandfather would drive him to the meetings and sit in wonder as we all geeked out. Caleb was the youngest one there, his grandpa the oldest.

Caleb had a Cupcake too, so we hit it off right away. It didn't take long to see he was very bright, self-motivated, articulate, and eager to learn. I remember wondering if he would consider working with me. As it turns out, he accepted my invitation to show up on the weekends. For such a young man, he's certainly gained some unusually rich experiences over the last year or so. We even met the president of Israel and explained 3D printing to him together!

With patience, I've discovered Caleb's specific talents and passions. Caleb loves to research and design. I try to gently provide clear goals and we work on the designs together. Hearing a fresh perspective is so valuable. I'm convinced the designs turn out stronger when we're both involved.

Watching young people discover and use this technology is one of the main motivating factors for me. I'm certain that 3D printers will be in every school soon. I launched my second Kickstarter campaign to promote giving 3D printers to local schools, but it was not successful. I remain focused on the goal — a printer in every school and every home. How will we do it? More slowly than I wish we could.

While I feel a real sense of urgency for America's schools and youth, it takes time to educate adults. Yes, adults. Kids "get" these printers almost immediately. The decision-makers in schools and in the lives of these young students are slower to learn and buy into the idea of acquiring a 3D printer. So, we at Printrbot, along with many of you, are on a slow, steady march to educate the young and not-so-young. The world is starting to take note. ◪

Brook Drumm is founder and CEO of Printrbot (printbot.com).

MARCHING ORDERS FOR A 3D PRINTING WORLD:

1. Buy a 3D printer of any kind and learn firsthand.

2. Constantly show it off — to neighbors, your kid's class, teachers, librarians, city officials, anyone with any influence or voice.

3. Start a family-friendly meetup or club. Get students involved if you want to spread interest quickly.

4. Give your printer purpose! Find a cause to put your printers to good use — fix things at school, print gifts for a charity, donate RepRap parts to local schools or interested students.

5. Be a 3D printing ambassador at your local school. Start by talking with the principal and students about what class or teacher is a good fit for this new technology. Prepare to go slowly and be led by the boots on the ground. Be open to their ideas and help any way you can. The first wave of early adopters will be tech-savvy teachers who see this new generation of makers already building momentum.

WE ARE CHANGING THE WORLD HERE, SO GET BUSY!

12 REASONS YOU SHOULD BE USING OPENSCAD

WRITTEN BY **TIM DEAGAN**

The greatest little 3D modeling program you (maybe) never heard of.

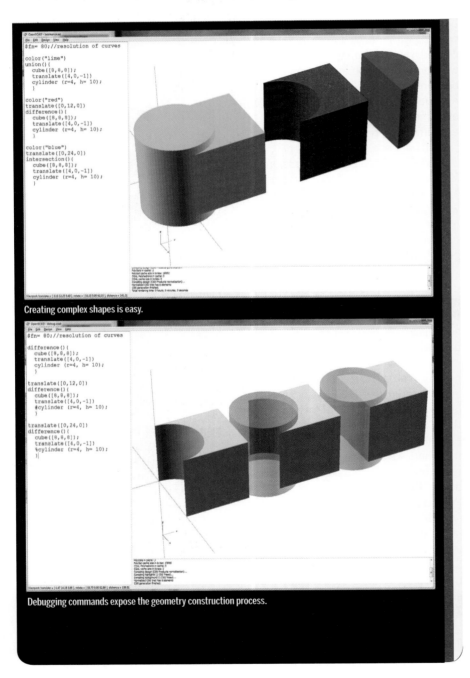

Creating complex shapes is easy.

Debugging commands expose the geometry construction process.

The biggest hurdle for 3D printing newbies is probably making original model files. Turning raw plastic into precision parts is mighty maker mojo, but how do you design the parts to begin with? The traditional approach to 3D modeling requires mastery of a complex graphical CAD interface before you can produce anything but the simplest shapes. If you'd rather skip all that and just type in the specifications of your design, Clifford Wolf and Marius Kintel's OpenSCAD modeling program may be for you. We think it's the best thing going for first-time 3D designers as well as seasoned pros. Here's why.

1. Free as in Speech, Free as in Beer

3D CAD tools can cost thousands of dollars. OpenSCAD is both free without cost (as in beer) and free to use, distribute or modify (as in speech) under the Gnu Public License. It's not just good for your wallet — it's good for your soul.

2. Cross-Platform Support

OpenSCAD works under OSX, Windows, BSD and multiple Linux distros. It runs well on older machines and doesn't demand high-performing video cards or drivers.

3. 3D Printing Pedigree

Many modeling programs are written with an eye toward animation or game design. The output looks great on screen but may not be printable. OpenSCAD was developed specifically for making physical objects.

4. Extremely Simple Interface

Unlike "virtual" 3D modeling programs, OpenSCAD creates models from text. There are no complex controls to learn, just three windows: one for status messages, another to edit text for the script, and a third to display the 3D results. The graphical window is used for viewing your model, not modifying it.

5. Easy To Learn

If you understand Cartesian coordinates, you're ready to start using OpenSCAD. `Cube([10,10,10]);` is a complete OpenSCAD script for a 10mm cube. `Translate([0,0,10]) Cube([10,10,10]);` moves that cube 10mm up the Z-axis.

6. Powerful

OpenSCAD uses constructive solid geometry to let you build complex shapes from simple ones: Combine two shapes to make one with the "union" operation, or subtract one from another with "difference." The resulting forms can be recombined using these (and other) operations indefinitely, to produce an infinite variety of shapes. Standard programming features like variables, loops, functions, and libraries allow you to extend the complexity as you grow your skills.

7. Tons of Example Files

OpenSCAD comes with the MCAD library, a collection of more than 170 importable modules that provide everything from double helix gears to Lego elements. Online, Thingiverse alone hosts more than 6,000 freely downloadable .SCAD files, with more added every day.

8. Geometry Debugging

Complex models can be frustrating to troubleshoot. OpenSCAD provides various settings designed to make the process of finding and squashing bugs much easier. Quick one-character operators set objects to background(%), debug(#), root(!), or disable(*) mode so you can focus in quickly on the part that's giving you problems.

A single line imports a model.

OpenSCAD is great for mash-ups. Stephen Colbert's head is a popular subject.

9. Rapidly View and Modify .STL Files

OpenSCAD imports .STL files regardless of what program created them and lets you interact with them just like "native" objects.

10. Change 2D to 3D

OpenSCAD can import 2D shapes in the common .DXF format and turn them into 3D parts using a process called "extrusion." Think pasta maker: Your 2D drawing is the die, and OpenSCAD lets you push dough through it to make a noodle of whatever shape you want.

11. Create Customizable Models On Thingiverse

Thingiverse's Makerbot Customizer automatically recognizes parameters in .SCAD files and provides a web interface so users can instantly change the size, count, internal dimensions or any other variable-defined feature in a model's script.

12. Strong Community Support

See OpenSCAD.org for a cheat sheet, manual, and list of online tutorials. Thingiverse users have assembled many impressive OpenSCAD collections that include amazing tools and extensions. Gary Crowell Sr.'s collection (makezine.com/go/garyscad) is a great place to start. ◪

Tim Deagan plays with fire, software, metal and motors in Austin, TX. Preferably all at the same time.

SOFTWARE FOR

Understanding your options each step of the way.

WRITTEN AND PHOTOGRAPHED BY **SEAN MICHAEL RAGAN**

CAD

CAM

CLIENT

RAKISH ANGLE

TOP HAT

BOW TIE

PEEPERS

SUPERHERO EMBLEM

There are basically four steps on the path from concept to printed object: the idea itself, the digital model, the tool paths, and the final print. Three layers of software — CAD, CAM, and "client" — bridge the gaps. For 3D printers, CAD and CAM (Computer-Aided Design/Manufacturing) refer to the production of digital models and their translation into physical instructions for the printer, while "client" software controls the hardware in real time.

Especially at the CAD level, there are many options, some of them quite expensive. Most of the programs in this list are at least free-as-in-beer, though many are supported by sales of more powerful "pro" versions.

Sean Michael Ragan is technical editor of MAKE.

3D PRINTING

MODELING / CAD

Even if you scan models from real objects, you'll probably want to adjust them in a CAD program. There are many file formats for 3D models, but almost all 3D printing CAM software expects STL. Unfortunately, not all STL files are printable. A printable model is "watertight," with a closed surface that clearly separates inside and outside. Computer graphics programs (like Blender) are usually "surface modeling" and don't care about watertightness. Newbies might first consider a more design-focused "solid modeling" tool (like 123D Design).

PROGRAM	DEVELOPER	SINCE	PRICE	NOTES
Blender	Blender Foundation	1999	Free	Renowned, powerful open-source surface-modeling program. Huge community. Steep learning curve.
SketchUp	Trimble	2000	$0/ $590 (Pro)	Good balance of usability and power. Built-in "3D Warehouse" model sharing feature has large community. No native STL support.
FreeCAD	Juergen Riegel, Werner Mayer	2002	Free	Very powerful, engineering-focused open-source parametric CAD platform. Feature set competitive with pro-line tools.
OpenSCAD	Clifford Wolf, Marius Kintel	2009	Free	Models are developed by textual scripting rather than virtual interaction. (See page 32.)
Sculptris	Tomas Pettersson / Pixologic	2010	$0	Uses "clay" metaphor for sculpting organic forms by "pinching," "smoothing," etc. Commercial use allowed.
Tinkercad	Tinkercad, Inc. / Autodesk	2011	$0	Fun, friendly, web-based tool for learning or simple modeling.
123D Design	Autodesk	2011	$0	Baby cousin of Autodesk Inventor. Streamlined solid modeler for PC, Mac, web, and iPad.
123D Sculpt	Autodesk	2011	$0	Extends sculpting metaphor with touchscreen interface. Cloud export process. iPad only.
123D Creature	Autodesk	2013	$0	Character rigging plus body sculpting/painting in an intuitive tablet-based interface. iPad only.
Cubify Invent	3D Systems	2012	$49	Solid modeling designed with 3D printing in mind. Exports STL, but does not import. Windows only.
Cubify Sculpt	3D Systems	2013	$129	Sculpting-based modeler designed specifically for 3D printing. Windows only. Imports and exports STL.

SLICING / CAM

"Slicing" programs translate 3D models into physical instructions for the printing robot, which can be visualized as a tangle of "tool paths" the print head will follow to fill in the model's shape. Most output industry-standard G-code files.

PROGRAM	DEVELOPER	SINCE	PRICE	NOTES
Skeinforge	Enrique Perez	2009	Free	Many settings give great control, but fiddly interface. SFACT is popular "simple" frontend.
Slic3r	Alessandro Ranellucci	2011	Free	Has largely superseded Skeinforge as premiere slicing engine.
Cura	David Braam / Ultimaker	2012	Free	Integrated CAM/client for Ultimaker and some RepRap-type machines. Fast CuraEngine slicer runs as background process. Exports G-code.
KISSlicer	Jonathan Dummer	2012	$0/ $42	"Pro" version supports multi-extruder printing, other advanced features.
MakerWare	MakerBot	2013	$0	Integrated CAM/client for MakerBot printers. Choose between Skeinforge and MakerBot custom slicing engine. Exports G-code.

PRINTER CONTROL / CLIENT

The "client" is basically the printer's control panel. It sends CAM instructions and provides an interface to control printer functions. As CAM and client programs advance, they are increasingly being combined into single-interface printing environments. (See page 90.)

PROGRAM	DEVELOPER	SINCE	PRICE	NOTES
ReplicatorG	MakerBot	2008	Free	Original MakerBot printer client. Largely superseded by MakerWare.
Pronterface	Kliment Yanev	2011	Free	Best known of three utilities in popular "Printrun" suite. Requires Python. Fiddly installation.
Repetier-Host	Hot World Media GmbH	2011	Free	Requires Python. Auto-installers available.
Octoprint	Gina Häußge	2011	Free	Web-based printer interface offering "anywhere" control, monitoring, and G-code visualization.
Afinia 3D	Afinia	2012	$0	Integrated CAM/client for Afinia/Up printers. No export.

$0 = Available at no cost under the terms of a proprietary license, with limited access to source code.
FREE = Available at no cost under the terms of an open license, with unlimited access to source code.

PLASTICS FOR 3D PRINTING

WRITTEN BY **STUART DEUTSCH**

As 3D printing explodes, so do the options among printable plastics.

ABS and PLA are the go-to plastics for most consumer-grade 3D printers, but the market is heating up quickly, and the barrier to entry is not very high. New types of plastic, blends of plastic with various additives, and grades of plastic formulated specifically for 3D printing are appearing all the time. With simple tools, it's even possible to turn pellets or other plastic scrap into usable filament right in your own shop. Our handy chart will help you get a handle on what's out there at the moment.

PRO TIP:
Even if it's technically the same plastic, not all filament is extruded alike. If you find a product from a particular manufacturer that works well in your machine, stick with it!

WOOD, STONE, AND STEEL

New filler-blended plastics mimic traditional materials:

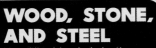

Designer/Dizingof

Laywood prints lighter at lower temperatures, and darker at higher.

Simon J. Oliver/Extrudable.me

Laybrick prints are heavy, with a surface like hard plaster.

Faberdashery.co.uk

It's easy to see how Galaxy Blue metal flake got its name.

Jeffrey Braverman

FILAMENT FRENZY

Temp	Material		Description	Price
265°C +	PC	Polycarbonate (7 OTHER)	Strong, tough, impact-resistant, prints clear. High melt temp requires special extruders.	$40 / lb
240 – 250°C	NYLON	Aliphatic polyamide blend (7 OTHER)	Slippery and slightly pliable. Good for low-friction parts. Taulman 618 takes dye well. 645 is stronger.	$20 / lb (618), $30 / lb (645)
215 – 250°C	ABS	Acrylonitrile butadiene styrene (ABS)	Arguably the most common 3D printing plastic. Strong. Many colors available. Unpleasant odor during printing.	$20 / lb
215 – 240°C	BENDLAY	Modified ABS formula (?)	Highly flexible and clear with great layer adhesion. Moderate strength. Recyclability unknown.	$20 / lb
220 – 230°C	HIPS	High-impact polystyrene (6 PS)	Like ABS but dissolves cleanly in limonene. Cheap support material alternative to PVA.	$18/lb
212 – 224°C	PET	Polyethylene terephthalate (1 PETE)	100% recycled. No odor. High clarity. Manufactured from food-grade materials.	$30 / lb (Taulman T-glase)
160 – 220°C	PLA	Polylactic acid	Easy-printing. Plant-derived and biodegradable. Many colors available. Comes in rigid and rubbery grades.	$20 / lb (hard), $30 / lb (soft)
180 – 200°C	PVA	Polyvinyl Alcohol	Easily dissolved in cold water. Commonly used as support material.	$40 / lb
120 – 150°C	PCL	Poly-caprolactone	Easily home-mixed with additives to form blends. Generally available only as pellets.	$20 / lb (pellets)

BLENDS

Plastics in their "natural" state are often tan, white, off-white, or clear. The wide range of colors available in plastics like ABS and PLA is created by mixing colored powders into the polymer base. Such additives can impart other properties, too, like improved conductivity, natural wood texture, phosphorescence, and various other effects.

CONDUCTIVE
Carbomorph is a conductive blend of PCL and carbon. It can be used to print working capacative buttons / sensors.

COMPOSITE
Laywood is 40% wood dust in plastic. It looks and smells like wood. **Laybrick** (chalk in copolyester) is also available.

GLOW-IN-THE-DARK
Glow-in-the-dark filament is a blend of ABS or PLA and phosphorescent powders like strontium aluminate.

METAL FLAKE
Galaxy Blue is a translucent PLA from faberdashery.co.uk containing engineered aluminum flakes to suggest "stars."

WARNING:

There is little hard data on the health effects of fumes emitted by plastics during printing. Some (like ABS) smell downright nasty, others (like PLA) actually smell pretty good, and still others (like PCL) don't have much smell at all. No matter what, err on the side of caution: Never operate a printer without plenty of ventilation.

Dr. Stuart Deutsch is a materials consultant in the NYC area and editor-in-chief at toolguyd.com.

JIVING WITH JARVIS

INTERVIEWED BY **GOLI MOHAMMADI**

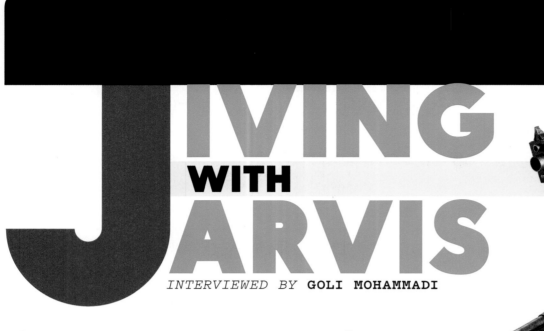

What would you do with a professional 3D printer?

Petaluma, Calif.-based artist Shawn Thorsson creates costumes and props of epic detail and proportion — we even featured him on the cover of MAKE Volume 32 alongside his 8-foot-tall Imperial Space Marine. Until recently, he predominantly used traditional mold-making techniques and tools for his builds. In the summer of 2012, he entered the Instructables Make It Real contest and won the grand prize of a $50,000 Objet30 3D printer, which he's named Jarvis. We chatted with him to find out how having a professional-grade printer at his disposal has affected his art.

Were you in the market for a 3D printer before entering the Instructables contest?
It would make more sense to say I was *watching* the market for 3D printers. I decided I needed one as soon as I'd first heard of it, but with prices getting lower and capabilities getting better almost daily, it made more sense to wait until I could get a machine that did everything I needed for a price I could afford. The Instructables prize just catapulted me ahead a few years.

How did you envision the printer was going to help you in your studio?
Mostly I figured it'd be like having another pair of hands. When I'm making details for props or costumes, I often have to cobble together all sorts of small parts out of whatever found items I have around. Rapid prototyping would allow me to have exactly the right thing instead of having to come up with a "close enough" thing.

What is your relationship with Jarvis like?
I love that machine. It's reliable and low-maintenance and continues to perform exactly as advertised — knock on wood. That said, I hate that machine. I used to feel that there was a lot more art to my various projects. Now, whenever I run into a sticky part of a build, something that used to require a creative, tangible solution, it seems like the easy answer is almost always, "I'll just have Jarvis make it." Before much longer I'm going to forget how to use that drill press thing collecting dust in the corner of my workshop.

What types of jobs is Jarvis best suited for in your studio?
I use him for prototyping small, highly detailed parts. If there's a widget that needs to be just so, I can feed him a model and count on it coming out perfectly while I work on other things.

Anywhere Jarvis fails?
There's a handful of things that he definitely can't do. Whenever I need a biological shape such as a creature's head or one of my garden gnome sculpts, it's almost always faster and definitely easier to sculpt out the prototype by hand using a lump of clay and some

"I love that machine ... that said, I hate that machine."

very basic tools.

The other thing I run into is problems with the printed material itself. The cured resin is a bit brittle. I dropped a prototype for a pistol prop and it shattered spectacularly when it hit the concrete floor in the workshop. It also has really poor memory characteristics. Thin pieces end up sagging under their own weight if they're not reinforced or supported in storage. Since I'm usually looking for a prototype to reproduce in more durable materials, this isn't a huge problem, but it definitely means that I have to start the molds pretty quickly after the printed parts are prepped, primed, and painted.

Still, the biggest problem I run into is people's perception. So many folks think that having this machine has given me a "magic button" solution to every problem. It's definitely not that.

Did you know how to make 3D files before owning a printer?

Yes and no. I have an engineering degree, which I earned back before the turn of the century. So I had some vague memories of CAD programs that I was trained on back when floppy discs were still in vogue. More recently, I spent a bit of time tinkering with 3D modeling when I added a low-end CNC machine to the workshop. His name is Lopez.

What CAD program are you using now?

I'm still a pretty low-budget operation, so my software suite is limited to the sorts of things that I can get for free. I've been making use of everything from SketchUp and Tinkercad to Blender and Netfabb. Each of them has strengths and weaknesses of their own, so I spend a lot of time importing and exporting files between them in order to take advantage of them all. I've learned by trial and error and the occasional online tutorial. It's safe to say I spend much time scrolling though help menus as well.

I'm also experimenting with trial versions of some more expensive programs because it's pretty clear that I'm not using my equipment to its fullest potential. Most of the time I end up bartering pieces and parts in exchange for digital models from folks who have greater skills than me. In short, ever since Jarvis was unwrapped, I feel like some kind of caveman using a smartphone to crack open nuts.

What plans do you have for Jarvis?

I don't have anything especially grandiose in mind. The first thing I had him working on was smaller detail parts for my own personal Ironman costume replica. I also had him make new weapon prototypes for my line of Combat Garden Gnomes. More recently, he's been vital for building essential parts for a hero costume I'm making for a short film that will be used to pitch a concept for a feature-length movie. The character will end up being in a practical suit half the time and completely CGI the rest of the time. Building the parts digitally with the 3D printer and CNC machine will make it almost impossible for the audience to tell the difference.

I've also got a few frivolous projects lined up. I had a friend of mine capture a 3D scan of my head. One of these days I'm going to make copies in various scales so I can make myself a Shawn Thorsson action figure. I'd also like to get a 3D scan of my skull so I can print it out and see what it looks like. Possibly mold it and make a pair of bookends, or a candy dish, or a paperweight. ◪

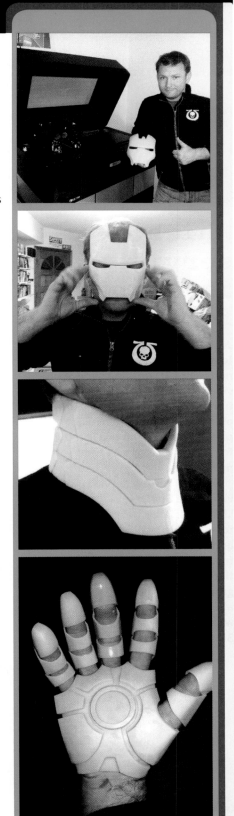

Shawn Thorsson

PRINTING WITHOUT A 3D PRINTER

WRITTEN BY **STETT HOLBROOK**

Which service is right for you?

Just as the field of 3D printers seems to grow by the day, so does the number of 3D printing service providers. There is a provider that will transform your ideas into a physical object, be it made from ABS or gold-plated brass. Which one is right for you? It depends on what your needs are.

Some of the big service providers like Shapeways and i.Materialise try to be all things to all people by offering services and features designed to appeal to a wide class of users. But as the 3D service provider industry grows so does market differentiation. Here are five makers, each with different reasons for using 3D printing services.

Katrien Herdewyn shared her sketches with a designer at Belgian 3D service provider i.Materialize, who produced the finished product.

David Collart

GoGoDynamo

Outsourcing through Shapeways allows Wayne Losey to get the quality necessary to make ModiBots pieces snap together properly.

Nervous

Nervous System's complex organic designs, such as this leaf structure-inspired Orbicular lamp, wouldn't be possible with traditional molding technology.

Design Services

Belgian fashion student, shoe designer and Ph.D. candidate in physics, Katrien Herdewyn is in her third year of school at the Academy of Fine Arts Sint-Niklaas (SASK). One of her assignments this year was to design a line of shoes using "now technology." For Herdewyn, 3D printing was the natural choice.

"You can make things you couldn't do with any other material," she says.

Trouble was she had very little experience with CAD. i.Materialise, a Belgian 3D service provider, offers design services to help customers like Herdewyn translate their ideas into printable objects. And many of their designers speak Dutch, Herdewyn's language of choice.

Herdewyn shared sketches of her shoes with a designer and they emailed back and forth for several weeks until one day her shoes arrived in the mail.

"It was very exciting," she says. "The box was very big, but very light. I was afraid it was going to come out with very tiny shoes."

Instead the shoes were exactly what she wanted. No adjustments necessary.

She was insistent that the shoes had

to be wearable, not something that all haute couture shoes achieve. She says the flexible, polyamide construction and open-air design make them quite functional as footwear.

"They survived a fashion show," she says.

Now that Herdewyn's got one round of 3D design under her belt, she's thinking of working with a 3D designer again to create her final collection of shoes next year.

Quality

Wayne Losey, co-founder of Dynamo Development Labs and creator of Modibots, is a former toy designer for Hasbro who started ModiBots to give kids and kid-like adults access to more than 400 interlocking 3D-printed parts that allow them to create the kinds of toys they want. It's the antithesis of an off-the-shelf product and perfect for the kind of on-demand manufacturing that 3D printing services allow.

While owning and maintaining a printer able to print out his SLS polyamide parts is cost-prohibitive, one of his chief reasons for outsourcing his prints is the quality he gets. ModiBots join together

with interlocking balls and sockets and that means the material must have a high degree of tolerance and flexibility. Outsourcing his work (he says he uses Shapeways because of their prices and good customer service) allows him to get the kind of quality necessary to make the toys snap together properly.

"SLS is the perfect material," he says. "It's a really good fit."

After settling on SLS, he now designs his parts with the material in mind — the material now drives the design as much as the design has driven the material.

Volume

Using a 3D service provider for a high-volume production run is not the norm. Small runs of made-to-order items that eliminate the need for inventory are one of the benefits of 3D printing. But as the public becomes familiar and interested in 3D designs, demand and therefore print volume will go up. Case in point: Nervous System.

Nervous System makes distinctive jewelry, lamps, and other home accessories. The organic and modern designs are complex and wouldn't be possible with traditional molding tech-

Steve Double

Sean Cusack's wolf head is made from several pieces of SLS printed steel TIG brazed together.

nology. Instead they use Shapeways to print in SLS, wax, and steel.

Jessica Rosecrantz, creative director and co-founder of Nervous system, says that while her primary reason for using a service provider is the access to high-end materials, printing in volume is another. As of this spring, the company had sold more than 30,000 pieces of 3D-printed jewelry since they opened shop in 2009. Many of those items were printed one at a time and on-demand, but with 60% of their profits coming from wholesale, a lot is printed in bulk and ordered in quantities as large as 300.

And yet the company strives to keep inventory for even the most popular items down to 100 items or less.

"We're a small business, and even if our designs were moldable, we don't have the finances or inclination to invest in a production run," says Rosecrantz. "We like the flexibility of 3D printing. We currently make hundreds of different designs and are constantly designing more. There's no startup cost to make any of them and no risk to put them out there."

Type of Material
Sean Cusack owns Sheet Metal Alchemist, a San Francisco-based company that does technical installation artwork, metalworking for furniture, architecture, sculpture, and interactive electronics. He says he uses 3D printing services primarily for the high-performance materials available and the ease of use.

"Pretty much all extruder-based technologies produce a surface finish which is unacceptable for any kind of final product that our company would make," he says.

He wants to make products out of materials that simply can't be made with a home- or shop-based printer — at least not one that most people could afford.

"The 'hot glue gun' look of melted PLA or ABS is not something that people are willing to pay big bucks for in a final product. Additionally, even as a prototyping platform, we run into trouble using extruder-based technologies because of the lack of support material present in the process. This means we are forced to design a part specifically for 3D printing which isn't very useful to us down the road.

"SLS, DMLS, and Polyjet technologies have served us well in certain situations. SLS and DMLS for producing metal and metal-ish parts with crazy geometries, without having to worry about where we will add filler material. We've actually printed a ~2x scale wolf head out of several pieces of SLS printed steel, and TIG brazed them together to make the

Cusack's wolf head up close.

full piece. It was pretty cool, but really expensive.

"From what I have been seeing from the installation-art realm, it looks like consumer 3D printers are primarily used to make small knickknacks or tchotchkes that end up as paper weights or pen holders. I'm excited to see the day when we can all afford our own SLS, DMLS, or Polyjet printers — some big gates will swing open, and some amazing things will be made!"

No 3D Printer

Utilizing a 3D service provider because you don't have a printer at home or in the shop is a pretty basic reason. But Steve Hoefer — maker, inventor, video documentarian and MAKE contributor — says he's got enough to keep him busy and he doesn't actually want a printer of his own. At least not at this stage of the industry. He says fiddling with a printer would take time away from his many hobbies.

"Right now owning a consumer-level 3D printer is like car ownership in the 19th century," he says. "Owning one is a hobby all by itself. Owners can spend as much time maintaining the machine as operating it. You can't use one without knowing complete details of how it works. Each machine behaves a little differently so there's no one guide that will help you achieve mastery. And even

when everything is working right, prints still go wrong. And of course that's exactly what makes them great machines for makers!"

Rather than start a new hobby, he wants a printing service to support his other hobbies.

"When I need a part I don't need to spend time testing the best way to avoid print warping or try several fill patterns and slicing algorithms to see which gives the best results," he says. "One answer would be to buy a professional level 3D printer, but they still run around $10,000 which is a little steep. The other answer is to use a service bureau like Shapeways or Ponoko. Time to get a finished part is much slower, but I don't have to spend any time dealing with printers, bad prints, or putting a nice finish on a print. They take care of all of that, and have more experience doing it than I could ever get."

The biggest downside, he says, is that iteration can be slow and expensive. Those 1-millimeter adjustments take time.

"So in short I think 3D printing is fantastic, and I'm glad to see it spread so widely. But right now it doesn't provide what I need, which is precision plastic bits with no hassle." ◪

Stett Holbrook is senior editor of MAKE.

COST COMPARISON

What does sending a design to a service provider cost? It depends on the material, of course, but also what provider you use. Here's a quick print comparison for Makey, MAKE's robot mascot, in various materials from five of the top service providers. The model stands about 2 inches tall.

SHAPEWAYS
shapeways.com
◼ **White polyamide:** $44.53
◻ **Glazed ceramic:** $19.31
◼ **Alumide:** $74.08
◼ **Gold plated glossy:** $340.78

I.MATERIALISE
i.materialise.com
◼ **White polyamide:** $16.59
◻ **Gloss white ceramic:** $19.30
◼ **Alumide:** $19.91
◼ **Brass:** $648.47

PONOKO
ponoko.com
◼ **White polyamide:** $101.01
◻ **Glazed white ceramic:** $46.45
◼ **Stainless steel:** $598.54
◼ **Gold plate:** $605.54

SCULPTEO
sculpteo.com
◼ **White polyamide:** $69.42
◻ **White glossy ceramic:** $26.51
◼ **Alumide:** $83.29
◼ **Stainless steel mirror finish:** $2,813.22

KRAFTWURX
kraftwurx.com
◼ **White polyamide:** $113.59
◻ **Fire glazed ceramic:** $28.96
◼ **Cast stainless steel:** $2,660.90
◼ **10K gold:** $17,052.80 (with gold selling for $1,400 an ounce, a hollow or "shelled" print would be cheaper.)

Fred Kahl

■ **TIME:** A Weekend ■ **COST:** $100–$300

3D SCANNING PARTY "PHOTO BOOTH"

Use this rig with a Kinect sensor to make full-body scans of your guests for 3D printing! *WRITTEN BY* **FRED KAHL**

When the 3D scanning software ReconstructMe came out last year, I was really excited about the idea of scanning and printing people at events and parties, but the deluge of office-chair-spin 3D prints that soon appeared online left me a bit disappointed. I wanted more!

Some people were doing full-body scans, but it involved painstaking work using mirrors or splicing together multiple meshes. I set out to devise a way to get a clean-scanned mesh that can be ready to print in minutes. The result: the Scan-O-Tron 3000. (Why 3000? Because it's better than 2000!)

This fast, full-body 3D scanning rig spins the subject on a heavy-duty turntable while a Kinect or PrimeSense sensor on a custom vertical rail captures their contours from head to toe. Pair it with ReconstructMe or Skanect software, and you can be spinning, scanning, and printing your party guests next weekend.

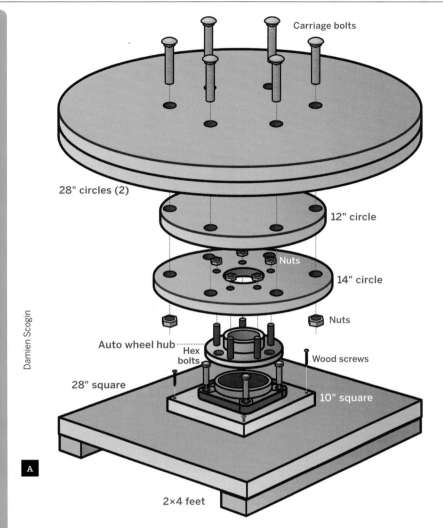

Carriage bolts

28" circles (2)

12" circle

Nuts

14" circle

Nuts

Auto wheel hub

Hex bolts

Wood screws

28" square

10" square

2×4 feet

Damien Scogin

A

Build the Heavy-Duty Turntable

Follow the assembly diagram (**Figure A**) and the instructions at makezine. com/projects/heavy-duty-turntable. It's pretty simple: Cut and drill the plywood parts, bolt them to a car wheel hub, and mount a high-torque rotisserie motor. 3D print the drive pulley, then hook up a belt made from an old bike inner tube. It works great. And I recently added a brake lever to hold the turntable still while people are stepping on and off.

Build the Scan-O-Tron 3000 Vertical Rail for Kinect Scanning

Follow the assembly diagram (**Figure B**, following page) and the instructions at makezine.com/projects/scan-o-tron-3000. First, 3D print the special carriage I designed to hold the Kinect. It's hinged and has a special "guillotine"

drop pin for locking it in an optimal downward angle.

Then build the vertical rail and frame out of MakerSlide aluminum extrusion, which is designed so that a wheeled carriage can run smoothly along it. Fabricate the scanner mount plate, then scavenge an old filament reel, a few feet of string, a couple pulleys, and some skateboard bearings to build the vertical positioning system.

Scan and Print Your Party Guests

Time to start scanning! ReconstructMe used to be the only game in town, which meant Mac users were left out or had to jump through hoops installing Windows under Bootcamp. Fortunately, that's all changed with the release of Skanect (skanect.manctl.com). Though I still use ReconstructMe, Skanect has several features that make it quite attractive. First off, it's

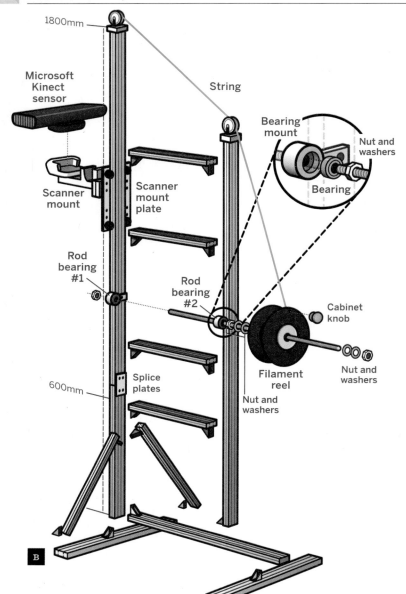

1800mm

Microsoft Kinect sensor

String

Scanner mount

Scanner mount plate

Bearing mount

Nut and washers

Bearing

Rod bearing #1

Rod bearing #2

Cabinet knob

Nut and washers

600mm

Splice plates

Filament reel

Nut and washers

Nut and washers

Damien Scogin

B

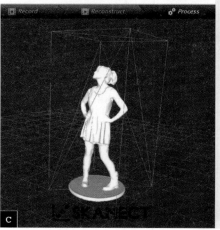

C

by highlighting the model green where you're getting good "point cloud" data and red where you're missing details. When the scan is complete, hit Record again to stop.

3. Reconstruct the Data
Skanect will process the point cloud and "reconstruct" it as a 3D mesh for you automatically. The Reconstruct tab is used to revisit saved scan files if you need to reconstruct them again. You can skip this tab for now.

4. Process Your Mesh
So now you've got a mesh scanned, but it's not 3D printable yet because it's just a skin and it's full of holes. It has to be converted to a solid volume. Click the Watertight button to close the holes and make it solid. Set Smoothing to Medium then click Run. The mesh will close and the full-color surface texture will appear. If you're printing on a desktop 3D printer, you won't need this texture — use the Remove Colors tool to see the actual geometry (**Figure C**).

cross-platform, offering both Mac and Windows versions. It's easy to install and offers basic tools to clean up your file and prepare it for 3D printing.

A word of warning, though — both programs require a newer machine with a serious graphics card. It was simple to download and install Skanect on my Mac, but I found that I needed to install the latest Nvidia CUDA drivers (nvidia. com/object/mac-driver-archive.html) to get GPU acceleration while scanning.

Launch the application and you'll see a sequence of 5 guided "step" tabs across the top of the screen to walk you through the scanning process.

1. Prepare a New Scan
Click the New button and you'll see that

you have controls to define the bounding box volume for the scan volume. For full-body scans, try 1m×1m×2.5m tall.

Before proceeding, you should also review the Settings. I found I had to lower the feedback quality settings to Medium or Low, or I would lose tracking while scanning.

2. Record Your Scan
Hit the Record button and start scanning (**Figure D**). There will be a little trial and error to find the right distance between the Kinect and turntable, and the right height at which to start scanning. At your starting point, you want to capture the top of the turntable in your scan volume, but not the floor.

Skanect gives you visual feedback

SKANECT VS. RECONSTRUCTME

SKANECT	RECONSTRUCTME
Free trial / 99€ ($130)	Free trial / 179€ ($235)
Mac and Windows	Windows only
Easy installation	Tricky installation with required drivers, but great support forums and helpful staff.
Easy to use GUI interface for setting dimensions of the scan's bounding box volume.	Preset scan volumes, but you can edit *config.txt* files to customize. We found the 2-meter default Human High-Res volume was too short for taller subjects.
"Future proofs" your scans by also capturing surface texture color data.	No surface texture map captured.
Saves images and metadata to a folder structure, average 70MB–90MB	Saves scan data as OBJ or PLY files, average 12MB–36MB
Integrated cleanup tools to make watertight meshes and planar bottom cuts, though the interface is a bit clunky (models are rotated with a slider that lacks precise control). It's still necessary to scale the model for 3D printing using other software, e.g. Netfabb. If they'd add a scaling feature, Skanect could be a one-stop solution.	No integrated cleanup tools other than reducing polygons (decimation), which I rarely use because it reduces print quality. Still, I prefer using ReconstructMe and Netfabb, since I've got to use Netfabb in any case for cleanup before printing.
Static scan display.	I love the final display of a scan — give it a swipe and it spins forever onscreen with a metallic silver surface.

The Simplify tool gives you the option of "decimating" the model — reducing the number of polygonal faces — to make it faster and easier to print. I typically bring it down to 100,000–150,000 faces for a human figure, but you can adjust to your liking. After you're done you can hit Colorize to reapply the texture if you want to store the color data with your model to "future proof" it.

Finally, perform a planar cut on the bottom of your model. Use the Move & Crop tool to rotate the scan on the x-, y-, or z-axis so the base is parallel to the ground. Use the Transform Z tool to position the cutting plane, then hit the Crop to Ground button.

5. Share or Export Your Model

Skanect lets you share your model to Sketchfab or 3D print it online at Sculpteo, but here we'll just use the Export feature. Export your file in STL format and you're almost ready to print.

Scale the mesh and print

The one issue Skanect doesn't address is scaling. Your model is still life-size, so figures will be 6 feet tall, not the 6 inches your printer expects. Open your STL file in Netfabb (free from netfabb.com), scale it to fit your printer, and execute an Auto Repair before resaving. You can watch how I do it at makezine.com/go/cleanup.

Load your STL file into the slicing program of your choice. I usually print in ABS with 3 perimeters and about 18% infill.

That's all you need to know to set up your own 3D scanning booth. Happy scanning! ▪

➕ See Kahl's World Maker Faire crew scans at makezine.com/go/wmfscans

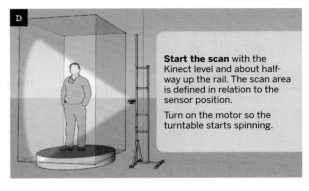

Start the scan with the Kinect level and about halfway up the rail. The scan area is defined in relation to the sensor position.

Turn on the motor so the turntable starts spinning.

Angle the Kinect downward on the hinge so the guillotine pin falls into place keeping it pointed about 30° down.

This ensures that the Kinect won't scan up pant legs or dresses and will capture the top of your subject's head.

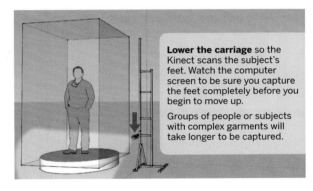

Lower the carriage so the Kinect scans the subject's feet. Watch the computer screen to be sure you capture the feet completely before you begin to move up.

Groups of people or subjects with complex garments will take longer to be captured.

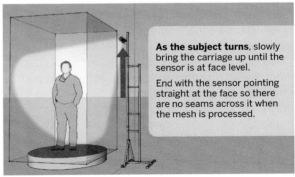

As the subject turns, slowly bring the carriage up until the sensor is at face level.

End with the sensor pointing straight at the face so there are no seams across it when the mesh is processed.

Fred Kahl, aka the Great Fredini, is an artist, sword swallower, world's worst magician, and the creative director of interactivity and gaming at New York's famed design studio Funny Garbage. He successfully Kickstarted his Coney Island Scan-A-Rama 3D portrait studio, where he scans performers and the public to populate his 3D-printed reconstruction of Coney Island's Luna Park as it stood 100 years ago.

METAL MADNESS

Move past plastic — use your 3D printer to cast objects in metal.

WRITTEN BY **MATT STULTZ**

Cosmo Wenman/cosmowenman.com

The ever-growing variety of 3D printing filaments is amazing — we can now print in strong, flexible, glowing, and dissolvable plastics — but sometimes plastic isn't the right material for the job. Sometimes, you just need metal!

Although the big boys in the industrial world can directly 3D print metal parts with laser sintering machines, this technology hasn't yet reached consumers. However, you *can* make your own metal parts at home with the help of your 3D printer and these easy-to-learn metal casting techniques.

ABS Molds for Bismuth Alloys

California artist Cosmo Wenman posted a tutorial on how to cast bismuth alloy parts directly into 3D-printed ABS molds (thingiverse. com/thing:26500). Bismuth alloys (makezine.com/go/bismuth) have a lower melting point than bismuth

NOTE: Many bismuth alloys contain toxic elements such as lead. Acetone is not a friendly substance either. Proceed with caution and follow proper safety measures when working with these materials. For a nontoxic bismuth alloy, try Field's metal (*see "Desktop Foundry," MAKE Volume 35, makezine.com/go/foundry*).

alone (~212°F versus 520.6°F). This low melting point means a hollow mold can be printed in ABS and it won't melt or deform when the molten metal is poured into it. After the metal has cooled, the mold can be split away or, for more complex objects, immersed in acetone to dissolve or soften the ABS so it can be easily removed, leaving behind only the final cast metal piece.

Some users are also experimenting with making two-part molds printed in plastics with higher melting points, like nylon. This allows them to cast alloys with higher melting points, like pewter.

Lost-PLA Casting of Aluminum

A common technique in jewelry making and manufacturing is *lost-wax casting* or *investment casting*. A model or "pattern" is made in wax, then a plaster mold is made around the wax model. When the mold is fired in a kiln, the wax is burnt out or "lost"; then metal parts can be cast in the mold.

This same technique can be used with PLA filament. Jeshua Lacock of Boise, Idaho, used the lost-PLA technique and a homemade furnace to cast aluminum parts for his home-built CO_2 cutting laser, going from design to print to metal part in just one day. He thoroughly documented the process at 3Dtopo.com/lostPLA.

Backyard furnaces fueled by charcoal or propane can get hot enough to melt aluminum (1,220°F) and bronze (1,742°F). Engineers at Coreprint Patterns in Hamilton, Ontario, even used the lost-PLA method to cast stainless steel (2,750°F), taking their mold to a local foundry that could attain the higher temperatures needed (coreprintpatterns.com/lost-pla).

Sand-Casting Bronze from 3D-Printed Patterns

When QC Co-Lab (qccolab.com) opened the doors on their new hackerspace in Davenport, Iowa, they wanted to celebrate in style. They used their 3D printer to create commemorative medallions — and then cast the medallions in bronze using a homemade charcoal furnace and the *sand casting* method. There's a great tutorial at Foundry101.com (foundry101.com/new_page_7.htm).

With these techniques, your personal desktop factory can help you create precious metal jewelry, aluminum parts for your robots, or bronze busts from your Kinect scans. The next time someone asks you, "Can you make metal parts with that thing?" you can happily proclaim, "Yes!" ◾

Matt Stultz is a leader of the 3D Printing Providence group (3dppvd.org), a HackPittsburg founder, and a MakerBot alumnus with experience in multimaterial printing and advanced materials. He was a member of this year's MAKE 3D Printer Shootout test team.

DIY FILAMENT EXTRUDER

■ **TIME:** A Weekend ■ **DIFFICULTY:** *Intermediate* ■ **COST:** $350–$650

Build a Filabot Wee and make all the plastic you can print, for a fraction of the retail price. *WRITTEN BY* **TYLER McNANEY**

FILABOT WEE SPECIFICATIONS

■ Maximum temperature: 350°C
■ Extrude rate: 5ipm–20ipm, depending on plastic and diameter
■ Feed screw speed: 35rpm
■ Input power: 120V AC or 220V AC
■ Power draw: ~300W
■ Footprint: 17"×7"×8"

TOOLS

For the kit:
» **Screwdrivers, Phillips and flat-head**
» **Wrenches, ⁷⁄₁₆" (2)**
» **Adjustable wrench**
» **Clamp**
» **Wire cutter/stripper**
» **Pliers, vise-grip**
» **Pliers, needlenose**
» **Allen wrenches, metric**
» **Hammer, small**

Additional, for the scratch build:
» **Metal lathe and/or mill**
» **Drill press and metal drilling bits**
» **Thread tapping set**
» **Plasma cutter, metal laser cutter, or water jet**
» **Bending brake**
» **Saws** for cutting plywood: jigsaw, scroll saw, table saw
» **Welding equipment**

Download the schematic diagram and PDF templates for all plywood and metal fabricated parts: makezine.com/ projects/diy-filament-extruder

Jeffrey Braverman

Is your 3D printer burning through filament? Guess what — there's no reason it has to be expensive. With a filament extruder, you can make your own filament from plastic pellets costing as little as $3–$5 a pound, saving you up to 90% compared to purchasing filament from online sellers.

We launched Filabot on Kickstarter with the goal of recycling waste plastic into something useful. The project has evolved into a family of machines, the first of which is our new Filabot Wee, which melts inexpensive PLA or ABS pellets and extrudes high-quality filament in 1.75mm and 3mm diameters.

The Filabot Wee is designed to use as few parts as possible while still making quality filament, quickly, for fused-filament fabrication (FFF) 3D printers. We sell it fully assembled, or as a complete kit. In addition, the Wee is an open hardware project under the terms of the BY-NC-SA Creative Commons license, and we're making our plans freely available so that anyone can build their own. Here's how.

BUILD YOUR FILA-MENT EXTRUDER

1. Check your parts

If you bought a kit, check the Bill of Materials to make sure all the necessary parts are provided, and contact Filabot if anything's missing.

1a. Make your parts (optional)

Scratch builders, you've got some wood- and metalworking in your future. To get started, download the PDF parts drawings from makezine.com/projects/diy-filament-extruder.

Using a laser cutter or a fine woodworking saw, cut the enclosure's top control panel and back wall from the 6mm plywood. Using heavier saws, cut the ½" plywood to make the base and the side panels.

Cut, drill, and bend the 11-gauge sheet metal, per the drawings, to make the motor mount, back chamber support, front chamber support, and 2 side brackets. Cut and bend the hopper from 18-gauge sheet metal, and cut the thrust bearing plate from ¼" sheet metal. We cut our parts on a plasma cutter, and it's really the way to go; a local CNC service may be able to do this for you.

Now to the machining. Chuck the 1018 steel rod stock in a metal lathe or mill to cut the basic profiles for the motor coupler, shaft collar, and large nozzle. Then mill the hex edges on the large nozzle. Go back to your lathe and/or mill to cut the 1045 steel rod as needed to fabricate the chamber, and the aluminum rod as needed to fabricate the chamber surround (aka heater holder). Drill and tap threads on all parts as indicated by the drawings.

Finally, drill out the ⅜-16 hex bolts to make 4 nozzles: 1.75mm and 3mm for the 2 test nozzles, and 1.35mm and 2.5mm for the 2 "undersized" nozzles, which are the ones you'll actually use most of the time.

2. Attach the motor shaft coupler

With the parts in hand, you can move on to assembly. First, attach the motor coupler to the gearmotor shaft, using two of the M6-1mm cap screws. Tighten them with an Allen wrench.

3. Bolt the motor to the motor mount

Attach the U-shaped motor mount to the gearmotor using four ¼-20×½" bolts with washers. Thread these bolts in loosely for now; you'll tighten them completely later.

MATERIALS

Filabot Wee Kit $649 from filabot.com, includes all of the following parts. You can buy them separately for about $325–$350 total (see makezine.com/projects/diy-filament-extruder for the full list of vendors and part numbers), but the kit will save you a lot of shopping as well as cutting, bending, milling, and welding the parts.

Bought separately, you'll need:
- » **Plywood, 6mm:** 11"×7" (1), 6"×8½" (1)
- » **Plywood, ½":** 16"×6" (1), 16×9½" (2)
- » **Steel rod, 1018 or 1020 alloy:** 1¼"×5" **long** for machining the motor coupler, shaft collar, and large nozzle
- » **Steel rod, 1045 alloy,** 1.125" dia., 9" **length** for the chamber
- » **Aluminum rod, 2½" dia., 3.8" length** for chamber surround/heater holder
- » **Hex bolts, ⅜-16×½"** (4) McMaster-Carr #91309A619, mcmaster.com
- » **Sheet metal, 18 gauge, approx. 12"×12"** for the hopper
- » **Sheet metal, 11 gauge, approx. 12"×12"** for the motor mount, chamber supports, and side brackets
- » **Sheet metal, ¼" thick, approx. 4"×3"** for the thrust bearing plate
- » **Insulation, rigid mineral wool, high temperature, 2⅞" ID, 4½" OD, 4" length** McMaster #9364K17
- » **Gearmotor, 35rpm, 24V DC, 3.5A, ½" shaft** Filabot #200-0017
- » **Cartridge heater, 110V AC, 300W, ⅜" dia.** Filabot #200-0018; or for use outside the USA, 220V, 300W, 8mm dia., Filabot #200-0022
- » **Thrust ball bearing, ½"×1⁹⁄₃₂"×⅝"** VXB item #Kit8751, vxb.com
- » **Feed screw Filabot #200-0020** or get #200-0020-UF and cut it yourself
- » **Thermal paste** such as Arctic Silver 5
- » **Power cord** McMaster #71535K54
- » **Power inlet, C14 type** Filabot #300-0002
- » **Power supply, 24V DC 4.5A** Filabot #300-0003
- » **PID digital temperature controller, with K type probe** Filabot #300-0004
- » **Wire, 16 gauge, insulated in green, blue, red, and black** about 36" of each
- » **Rocker switches (3)** Filabot #300-0009
- » **Blade connectors (6)** Allied Electric #70083321, alliedelec.com
- » **Fork connectors (8)** McMaster #69145K72
- » **Terminal block, 5-pole** McMaster #7527K45
- » **Wire nuts** McMaster #7108K2
- » **Terminal jumpers (2)**
- » **Hex bolts, ¼-20:** ½" (8) and 1" (4)
- » **Hex nuts, ¼-20** (8)
- » **Flat washers, ¼"** (8)
- » **Cap screws, socket head, M6 (6mm) × 1mm × 12mm** (5)
- » **Setscrews ¼-20 × ⁵⁄₁₆"** (2)
- » **Wood screws, #4:** 1" (10), ½" (4)
- » **Finishing nails, 18 gauge × ¾"** (8)

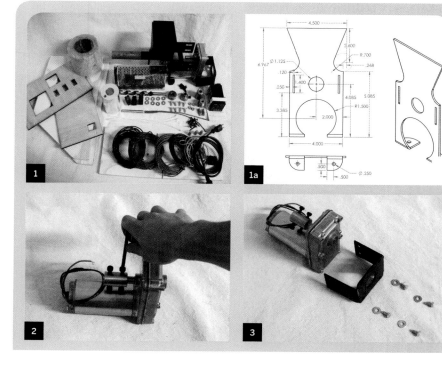

1

1a

2

3

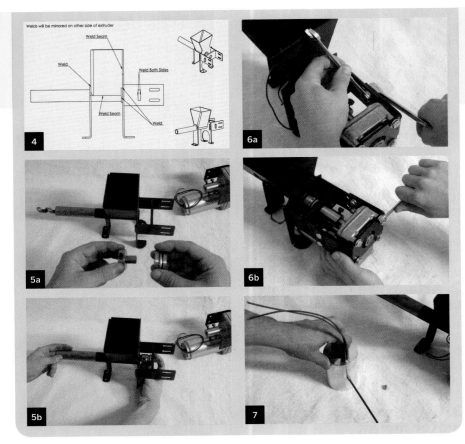

TIP: The motor mount may be spaced a bit away from the chassis; squeeze them together with vise-grips if needed to start the bolts.

7. Install the heater

Apply a small bead of thermal paste around the heater to help with heat transfer. Then slide the heater into the smaller lengthwise hole in the heater holder and tighten a setscrew just enough to hold it in place.

Apply thermal paste to the inside of the larger hole, then slide the heater holder onto the chamber with the heater wires sticking out the back. Insert a setscrew into the side hole as shown, and tighten.

8. Temperature probe and insulation

Screw the temperature probe into the top hole in the heater holder, using thermal paste to ensure a steady reading. Tighten with an adjustable wrench.

Zip-tie the insulation around the heater holder with the seam on top, and trim the zip ties.

Finally, screw the large nozzle into the front end of the chamber.

9. Mount the base

Fix the relay and power supply to the base using ½" wood screws as shown. Bolt the extruder to the base using ¼-20×1" bolts, washers, and hex nuts.

4. Weld the extruder chassis (optional)

If you're building from scratch or from the unwelded kit, follow the Extruder Welding Diagram to weld the chassis together from the chamber, the chamber supports, the thrust bearing plate, the side brackets, and the hopper.

5. Mount the feed screw and thrust bearing

Slide the feed screw into the extruder chamber from the front, until its shank peeks out the back. Insert the shaft collar into the thrust bearing as shown, then hold them inline with the feed screw and push the screw the rest of the way in. When it stops, the shank should stick out past the bearing about ¾". Tighten the collar's cap screw with an Allen wrench.

6. Mount the motor

Slide the motor coupler onto the feed screw, making sure that one of its cap screws lines up with the flat spot on the feed screw shank. Tighten the cap screws with an Allen wrench.

Use four ¼-20×½" hex bolts, 4 wash-ers, and 4 nuts to attach the motor mount to the extruder. Before the bolts are tight, pull back on the whole gear-motor to take up any slack in the thrust bearing. Then tighten the bolts with two $7/_{32}$" wrenches, one on each side.

Once these 4 bolts are in, cinch down the 4 motor bolts. Saving these for last helps with shaft alignment.

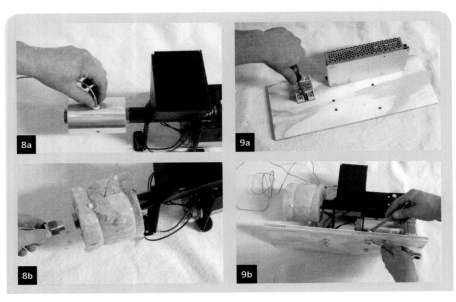

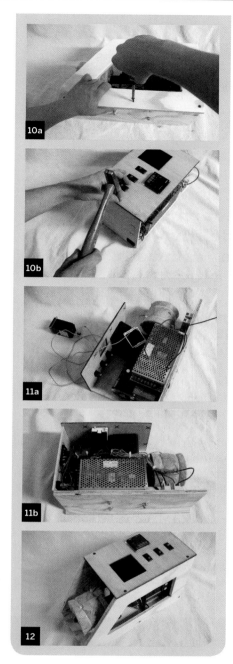

Position the rear wall (with the power inlet cutout and zip-tie holes) between the base and the top panel, and secure it with finishing nails through the top panel and side wall.

11. Wire the system

Follow the schematic at makezine.com/projects/diy-filament-extruder to wire the machine and mount the switches and power connector.

To connect the PID temperature controller, run its wires from inside the Filabot, through its white mounting frame, and out through the hole in the top panel. Connect the wires, then guide the temperature controller and mounting frame back into the hole.

Use 2 zip ties to mount the terminal strip to the rear wall, passing them through the precut holes.

12. Close it up

Install the left side wall with 3 screws through its face into the base, and 2 through the top panel into its edge.

NOW MAKE FILAMENT!

1. Place the machine on a tabletop, with its nozzle at the edge.

2. Turn all switches off. Plug in the Filabot, then turn on the Main Power and Temperature Controller switches. To use the temperature controller, follow the bundled manual. PID temperature controllers do some calculus to predict ups and downs and keep temperatures extremely smooth.

3. Set the temperature for the plastic you're extruding and wait until the chamber reaches the desired temperature. Only extrude at the proper temperature; running the machine too cold can damage mechanical components.

ABS extrusion temperatures range from 150°C to 180°C. PLA extrusion temperatures range from 130°C to 160°C. The temperature controller measurements will read slightly lower than the actual extrusion temperature. Adjust as needed for performance.

4. Put pellets in the hopper, and turn on the Feed switch. The first time you run the extruder, the first ½lb of filament or so may have a gray color from the oil used during machining. We don't recommend printing with it until the filament comes out clean.

As the filament starts to come out, guide it onto the floor below and lay the first few feet in a circle. This will make the filament naturally coil on the floor. From this point, just keep the hopper topped off and filament will keep pumping out. Every so often, check the filament for air bubbles. If they appear, lower the temperature as needed to eliminate them.

5. Once you're done making a batch of filament, turn all the switches off, unplug the power cord, and leave it unplugged until the next usage.

The Filabot Wee has been tested with ABS and PLA, and can make both 1.75mm and 3mm-diameter filament. Most of the time, you'll use the "undersized" nozzles; these are drilled smaller to compensate for plastic's tendency to swell during extrusion. The test nozzles drilled at exactly 1.75mm and 3mm are included so that if you're extruding a different grade or type of plastic, you'll be able to measure the swelling and make your own nozzles to compensate.

CAUTION: The Filabot may work with other plastics, but if you experiment, be sure you understand any potential hazards. PVC, for instance, may emit toxic fumes during extrusion.

You can rig your own reel or other method of collecting fresh filament as it extrudes, or try our Filamatic spooling kit. It has a 12rpm gearmotor and a custom circuit board and uses 2 photo gates as switches. It works with both 1.75mm and 3mm filament diameters and can handle any extrusion rate. ◾

Tyler McNaney's childhood toys never survived unscathed by his flat-head screwdriver. He is still intrigued by how things work.

10. Build the enclosure

Turn the base on its side and attach the right side wall (with the big cutout) using three #4×1" wood screws through its face into the base edge. (This is the "right" side wall with respect to the front of the extruder.)

Flip the unit upright and use two 1" wood screws to mount the top panel to the right side wall.

TIP: Position the base ½" up from the side walls' bottom edges so they'll act as feet.

MORE FUN PROJECTS 3D

WRITTEN BY **CRAIG COUDEN**

Mikola Lysenko

Print figurines and toys, learn scanning and modeling tricks, and make an extruder from a diesel glow plug!

PRINT YOUR VIDEO GAME CHARACTER IN 3D

I'm a big fan of video game character toys (recently springing for a killer set of posable Orbital Frames from Zone of the Enders), but as characters become more customizable, 3D printing may be the better way to bring your personalized gaming persona into the real world. FigurePrints (figureprints.com) can export your battle-weary, Level 90 Paladin from World of Warcraft or your hand-built, palatial masterpiece from Minecraft and print them on Z Corp machines in all their full-color glory. MineToys (minetoys.com) will print a figurine with your Minecraft skin.

For the DIYer with more retro sensibilities, Mikola Lysenko posted an ingenious method for creating 3D models from 8-bit style characters. Old favorites Mario and Zelda can now face off against new favorites like Meat Boy and Gomez for the fate of your desk space! makezine.com/go/3d-8-bit

FigurePrints

PUSH PUPPETS

For a fun, quick project, update the classic push puppet toy, where you push the button and the figure collapses, then let it go and it springs back up. Thingiverse user Spencer Renosis shows how to design your puppet in Tinkercad, 3D print the parts, then add fishing line and a spring. makezine.com/go/3dp-push-puppet

MAKE IT STAND

Ever print something awesome, only to realize there's no way it can stand on its own? Researchers from the Interactive Geometry Lab at ETH Zurich University developed free software to optimize 3D models to balance in crazy ways. You select one or more base points, then the program alters the outer surface and inner mass to balance the object. The results are awesomely precarious. makezine. com/go/make-it-stand/

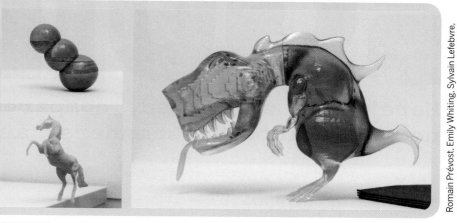

GLOW PLUG 3D EXTRUDER

There are very few, if any, open designs for a 3D printer hot-end and extruder that doesn't require 3D printed parts. Here's an exception. MAKE author Adam Kemp shows you how to build a high-performance extruder using only a handful of tools and parts, including a diesel glow plug from the auto parts store. makezine.com/projects/glow-plug-3d-printer-extrude

DIY PHOTOGRAMMETRY

Autodesk's 123D Catch is great for making 3D scans from digital photos, but it's a cloud-based freebie that limits the number of your photos and the resolution of the resulting 3D mesh. Digital compositor Jesse Spielman wrote an incredible tutorial showing how to use VisualSFM to roll your own point clouds without arbitrary limits, then use Meshlab to generate a mesh and paint it with the full-resolution photo texture. makezine. com/go/diy-photogrammetry

3D PRINTING BUYER'S GUIDE

WRITTEN
BY
ANNA
KAZIUNAS
FRANCE

Here's how we tested, compared, and rated 30 new 3D printers, scanners, and filament bots.

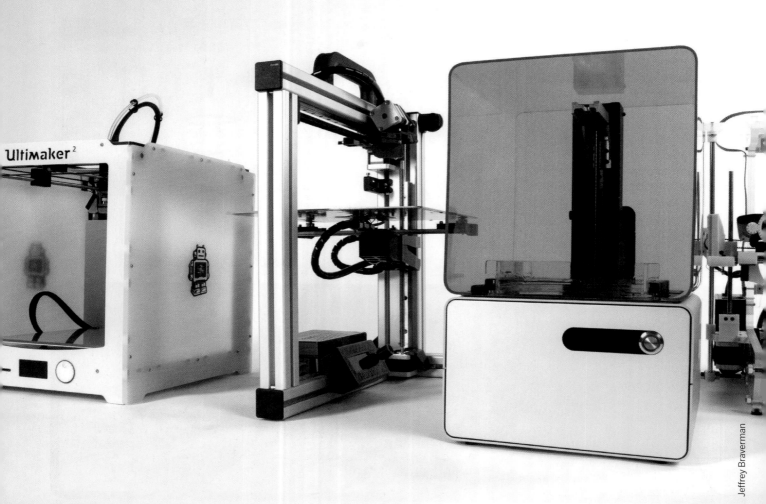

Which printer is right for you? The answer largely depends on your goals, budget, and user type. Will your machine be used at home in hobbyist pursuits and projects with the kids, or are you a designer or engineer who is considering a printer for your work? Must it accommodate the needs of multiple users in a school or makerspace? Is it a good value for the price? Are you a tinkerer or do you just want to hit the print button?

We thought carefully about this question as we prepared for our second shootout and decided that metrics alone wouldn't give us the complete answer. Instead, we opted to review this year's collection of printers with a more qualitative evaluation of the user experience. We increased the duration of our testing to include a setup period and shootout weekend, to better accommodate the learning curve that comes with every printer. We also focused our testing protocol on what we could realistically evaluate with a team of experts in the time allotted. We tested more machines than ever (adding 3D scanners and filament-making machines as well). Although we were unable to cover every printer on the market, we solicited fully assembled machines from every manufacturer that we were aware of, and if they sent us a machine, we tested it. Ultimately, we wanted to give readers a clear picture of what each machine does well, what it does poorly, and who it would serve best.

Are you ready to start printing? Then dive in and find the right tool. The options abound.

This Year's Trends in 3D Printing

Developments in desktop 3D printers come at a breakneck pace, and this year's roundup brought exciting new trends emerging in size, shapes, and print materials.

■ SMALL AND AFFORDABLE
Until recently, there has been a steep drop off when entering the small, sub-$1,000 3D printer realm. But with a couple new machines in our test, we now have access to printers that aren't just portable and low-cost, they're also capable of producing respectable results.

■ PRECISE PRINTS WITH LIQUID RESIN
Light-cured resin printers, new to our testing, are in a class by themselves, both in terms of the SLA technology used and the quality of the prints produced. Both of the machines we tested in this category are quickly being adopted by designers and craftspersons who require professional-grade details.

■ THREE-ARMED ROBOTS AND AUTOMATIC ADJUSTMENTS
With smooth, fast-moving heads, delta robot printers also entered the 3D printing conversation this year; we tested the OpenBeam Mini Kossel while keeping our eye on nearly a half dozen more (see our *"Ones to Watch"* section on page 100).

The Mini Kossel and the more typical Up Plus 2 also introduced auto-leveling features, which eliminates one of the monotonous manual steps required when setting up a print job.

■ THE PROSUMER CATEGORY GROWS
Until recently, the prosumer selection for fused-filament printers has largely been limited to the MakerBot Replicator 2. However, the new Ultimaker 2 now provides a viable option to those needing high-quality results, onboard controls and intuitive high quality software, with a form factor that is suitable for a professional workplace.

MEET THE TEAM

WE ASSEMBLED AN AWESOME CREW FROM ACROSS THE COUNTRY, INCLUDING:

- Leaders of 3D printer build groups
- Superusers of different printers
- Designers
- Artists
- Teachers of design and technology
- New users who don't have printers
- Engineering students
- Makerspace 3D printing trainers and operators

ANNA KAZIUNAS FRANCE led this year's 3D printer shootout and coordinated MAKE's reviews of printers, scanners, and more. She's director of the Providence Fab Academy, dean of students for Fab Academy Global, co-author of *Getting Started with MakerBot*, and editor of the forthcoming *Make: 3D Printing* book. You can find her online at kaziunas.com.

KACIE HULTGREN is a Thingiverse superstar and a scenic designer in the theater industry, using her MakerBot to make amazing furniture and details for scale models. She's also an online instructor in "Design for Desktop 3D Printing" at Maker Training Camp. Find her creations at prettysmallthings.com.

ANDERSON TA is a digital fabrication expert. By day he oversees the dFab Studio at the Maryland Institute College of Art. By night, he operates Matterfy LLC, promoting and evangelizing 3D printing hardware. He's led some of the very first 3D printer workshops in the USA. Chances are good that you've run into him at the many Maker Faires around the country.

MATT GRIFFIN, the leader of last year's shootout, is director of community and support at Adafruit Industries, a former MakerBot community manager, and author of the forthcoming book

Design and Modeling for 3D Printing.

JOHN ABELLA, an obsessive hobbyist and 3D printer enthusiast, has run 3D Printer Village at World Maker Faire New York since 2010 and wrote for the first *MAKE Ultimate Guide to 3D Printing*. He's currently preparing to teach 3D printer assembly workshops with BotBuilder.net.

ERIC CHU, MAKE Labs engineering intern and resident 3D printer guru, is also an engineering student, yo-yo hacker, robot builder, and fried rice aficionado.

MATT STULTZ is the leader of the 3D Printing Providence group, founder of HackPittsburgh, and a MakerBot alum, with experience in multimaterial printing and advanced materials.

TOM BURTONWOOD is an artist and educator at Columbia College and the School of Art Institute of Chicago. He is co-founder of Chicago's The 3D Printer Experience hybrid makerspace and retail store. His work has been exhibited from Brooklyn to Miami to Los Angeles to Osaka and can be seen at tomburtonwood.com. Tom also co-founded Improbable Objects and participated in the MakerBot MET#3D Hackathon at New York's Metropolitan Museum of Art.

LYRA LEVIN is a climber, aerialist, contortionist, parkour noob, and Ninja 500 rider. She is a compulsive builder of things and member of industrial arts collective Ardent Heavy Industries.

CHRIS MCCOY is a 3D printing instructor at TechShop San Francisco. He is also co-founder of You3Dit, a distributed global network of 3D printers and CAD designers.

BLAKE MALOOF is a game designer at Toys for Bob (*Skylanders*), and authored the Tinkercad tutorial in the first *MAKE Ultimate Guide to 3D Printing*.

ERIC WEINHOFFER is the Maker Shed product development engineer. He's been an owner-operator of 3D printers since interning at MakerBot in 2009.

BEN LANCASTER is an engineering student and MAKE Labs alum.

JAMES CHRISTIANSON produces video for MAKE and does 3D modeling and animation on the side. He is working with an eight-person team to build a prototype 3D-printed Iron Man suit.

DEREK POARCH is a computer engineer with emphasis in mechatronics and embedded systems. He is creating his own large-build 3D printer.

Gunther Kirsch

THE CHALLENGE PRINTS
You Can Learn a Lot from a Robot

We used Makey, the MAKE robot by Eric Chu, as our comparison test print (thingiverse.com/thing:40212). It was carefully chosen for the following reasons:

The tops of the hands and head are formed from **small circular layers**. These show how finely the extruder is able to handle short extrusions.

"Ringing" imperfections on the surface of a part are essentially the "ghosting" or repeating of features, caused by vibrations induced when the nozzle rapidly changes directions. It happens mostly with the M logo on the chest, and it's a test of the machine's overall stiffness and the acceleration settings in its software and firmware.

Bridging where the two legs meet the body shows how well the slicing software is calibrated. There should be little sagging where the plastic bridges between the legs.

Multiple **"islands"** require the extruder to retract the filament and move quickly to the next island. A good extruder with good retraction settings should produce prints with no "stringing."

Fine details in the logos on the chest and back show off how well the slicing software and hardware handle tiny features.

Slight **overhangs** test how well the machine cools the plastic — it'll sag if there's too much heat buildup.

The robot's head also shows how well the slicer handles **domed/spherical surfaces**. There should be no gaps or holes.

Concave and convex curves are a good test of backlash caused by untensioned belts. An axis can't move until the backlash catches up with the turning of the pulley — so instead of two axes moving at once to form a curve, only one moves and creates a flat spot.

Straight, tapered, and curved walls of the legs, body, and arms should be smooth — if not, it's a sign of "Z wobble," when the z-axis doesn't travel straight up or doesn't move the same distance after each layer, often due to low-quality or off-center leadscrews.

James Burke

Preassembled Secret Heart Box by emmett

◗ This model tests a printer's ability to create complex, preassembled mechanisms.

◗ thingiverse.com/thing:44579

Spiral Lightbulb Sculpture by benglish

◗ This print was our torture test, designed to push and show the limits of each machine.

◗ thingiverse.com/thing:12108

Zombie Hunter Head by Sculptor

◗ We printed this as big as possible on each machine to show surface finish and see how each machine handled long prints.

◗ thingiverse.com/thing:69709

Goldberg Polyhedron 8,3 (1937) by George Hart

◗ This delicate mathematical structure shows off the extremely high resolution of the liquid resin printers in our test.

◗ georgehart.com/rp/rp.html

PRINTRBOT SIMPLE

PRINTRBOT / PRINTRBOT.COM

This surprising little printer is perfect for the classroom or anyone on the go.

WRITTEN BY **TOM BURTONWOOD**

Price as tested $399 assembled

Print volume 3.9"×3.9"×3.9"

Heated bed? No

Print materials PLA

OS supported Linux, Mac, Windows

Print untethered? With SD card, initiated from computer

Open-source hardware? Yes, noncommercial

Open-source software? Yes

Printer control software Repetier-Host / Printrun

Slicing software Slic3r

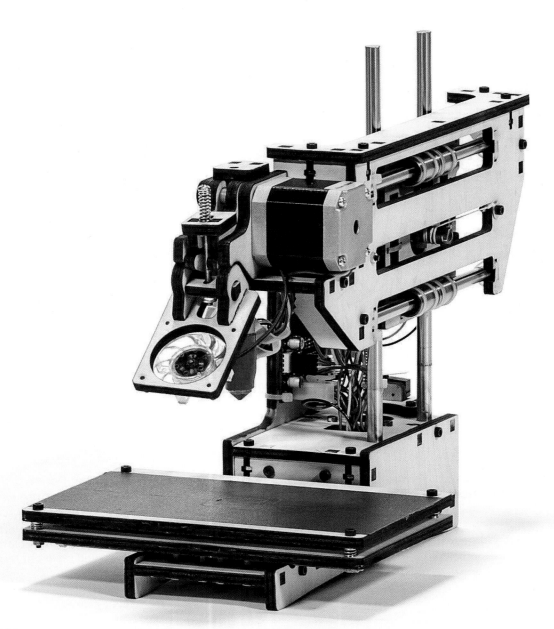

Gregory Hayes

It's easy to fall in love with the Printrbot Simple, one of the smallest, cheapest, most straightforward 3D printers available. For an entry-level maker, educator on a budget, student learner, or weekend warrior, this is a perfect machine in terms of cost, ease of use, and results. While most manufacturers try to cram new features and bloat into their products, Printrbot has stripped away as much as possible in the Simple.

The build area of the Simple is roughly 4 inches square. It's a perfect size for students of 3D printing everywhere, offering plenty of volume for making pint-sized components, action figures, and more, with enough constraint to keep it challenging.

$399 Assembled, With Open Design Files

We tested the assembled model ($399), but the Simple is also available as a $299 kit with all the parts you need for a complete build, or for the maker with access to a laser cutter, a $259 internal components-only option. The really cool part is that all their designs are shared online using a Creative Commons Attribution-Noncommercial ShareAlike license. This means you can produce your own Printrbots from scratch — you just can't sell them.

Setting up the Simple and getting your first print off it is pretty straightforward using the included printed manual or PDF from their website. It includes screen shots for both Mac and Windows operating systems. For the beginner, the troubleshooting guide is a bit lacking — more documentation explaining how to adjust the Z height, level the bed, and load filament would be useful.

Prints Untethered (But Needs Documentation)

We were really excited to learn that the Simple could print from an SD card, rather than requiring a laptop be tied up when doing an all-day print. However the lack of documentation from Printrbot on this matter was problematic, as was the way the Repetier-Host software seemed to handle our attempts (a bug with the application, not

with Printrbot's hardware). So we wrote our own instructions for new users on how to use the SD card for untethered printing (see page 87).

Other notes for first-timers: When the Simple starts a print, there's a fair amount of clicking and whirring, none of which seems to suggest anything is wrong. This could possibly be fixed with a C-clamp mounted to the back for extra weight or by securing it in a bench vise if you want to lose portability. There does seem to be a bit of droop on the extruder, too — be aware of this when leveling the build platform, and compensate if needed.

The bed is unheated and will only support PLA or nylon filaments. Early versions of the Simple did not come with an extruder fan or end stops; this has been corrected, and all printers now ship with these parts as standard.

We really like the Simple because of its lightweight platform and small footprint, making it easy to travel with to class, demonstrations, and meetups.

Good Prints, a Little on the Slow Side

The biggest drawback to the Simple has to be the slow speed at which it operates, but nonetheless with pleasing results. It did a pretty good job reproducing the smaller details of the MAKE robot, while a half-scale, 0.2mm layer height Zombie Hunter Head took about 2 to 2.5 hours with some really impressive results.

The Simple definitely kept up with the rest of the pack in terms of print quality and detail.

Conclusion

During testing, the MAKE team raved about this printer nonstop. (I liked it so much I bought one online before I left Sebastopol.)

For an educator, the Simple is perfect. Many students can afford to buy this kit themselves, solving the issue of capacity and hands-on learning in the classroom in one fell swoop. Building and operating a 3D printer are valuable skills for anybody looking to stay competitive in tomorrow's world, and the Printrbot Simple is a great place to start. ◪

I liked this printer so much I bought one online before I left Sebastopol.

HOW'D IT PRINT?

1.1875"

OPENBEAM MINI KOSSEL

OPENBEAM USA / STORE.OPENBEAMUSA.COM

The first deltabot in our tests printed well but it's still a machine for the hands-on tinkerer.

WRITTEN BY **ANNA KAZIUNAS FRANCE**

Price as tested $899 kit

Print volume 7"×5.9" dia.

Heated bed? No

Print materials PLA

OS supported Linux, Mac, Windows

Print untethered? Not yet

Onboard control interface Yes

Open-source hardware? Yes

Open-source software? Yes

Printer control software Repetier-Host

Slicing software KISSlicer

Gregory Hayes

Defeat the Cartesianists! Unique in our tests this year, the Mini Kossel's "deltabot" design has three arms, each attached to an independent, stepper-driven vertical axis, which results in a simpler machine with a smaller footprint and faster positioning than other printers.

Deltabot History

The Mini Kossel is an open-source parametric "delta robot" 3D printer, first built in Seattle in 2012 by Johann C. Rocholl, the father of deltabot printers.

In deltabots, linear motion is generated by three drive towers, so the print head can move equally fast in the x-, y- and z-axis, with fewer moving parts. A Bowden extruder keeps the head light and balanced. (The deltabot design also bypasses the Stratasys patent on heated build chambers, which specifically calls out a Cartesian platform.)

The OpenBeam Kossel family of printers is being developed at Metrix Create:Space in Seattle, led by Terence Tam, creator of OpenBeam aluminum extrusions. (OpenBeam just completed a successful Kickstarter for the Mini Kossel's big brother, the OpenBeam Kossel Pro, which won an Editor's Choice award at Maker Faire Bay Area in May 2013.)

Mini Kossel: Kit Only

The OpenBeam Mini Kossel we tested was a prototype, just the second of its kind. It's slated to be sold this fall as a mechanically complete kit, minus the 3D-printed parts, electronics, motors, build platform, and hot-end; these nonmechanical items will be available as configurable options. Depending on the electronics package chosen, the full deltabot kit will retail for $810–$899.

The machine was sent to us pre-assembled (at our request) and without instructions (which are in progress), so we're unable to rate the kit's documentation or ease of assembly. General documentation and a complete bill of materials for the Mini Kossel are available on the RepRap wiki at reprap.org/wiki/Kossel.

Design Highlights

The OpenBeam Kossel family is unique in the RepRap world as it maintains 100% part compatibility between Pro models (with injection-molded parts) and RepRap models (with 3D-printed parts). Both branches are open source.

The OpenBeam Mini Kossel also features improvements such as an automatically deployed bed leveling probe that compensates for bed tilt and bed height calibration.

By day, Tam is a mechanical design engineer for a microscope company, and it shows in the Mini Kossel's construction. He used affordable copies of high-grade commercial linear recirculating ball rails, like those in fine optical instruments, to achieve smooth linear motion and rigidity.

Nice Prints, Unique Auto-Leveler

Overall, we got very nice prints from the Mini Kossel. It printed the only MAKE robot with no "ringing" flaws (see page 59), and it also did quite well on the difficult Secret Heart Box model. The auto-leveling probe is fantastic; I thoroughly enjoyed watching it hop around the build platform before each print.

We experienced retraction issues when using the provided slicing configurations, leading to tiny overextruded "warts" on some prints. After consulting Tam and Rocholl, we were able to increase the retraction settings and dramatically decrease this problem.

Printing Untethered

We also experienced issues with the Vici LCD display and were unable to print untethered using onboard controls or SD card due to firmware incompatibilities. I asked Tam about it. "Debugging the SD card issue is one of our higher priorities," he said. "We're actually developing our own electronics, and it's our intention to have SD card printing enabled on our electronics."

Conclusion

This is not a printer for beginners, but if you're a tinkerer in search of something different and you're looking for your next kit build, we highly recommend the Mini Kossel. We had a fantastic time with it, and I'm seriously considering building one myself. ◪

Looking for a radically different 3D printer *and* a new kit build? The Mini Kossel is for you.

HOW'D IT PRINT?

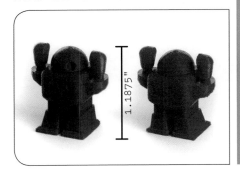

1.1875"

BUKITO

DEEZMAKER / DEEZMAKER.COM

With its small footprint and simplicity, it's the new standard for portable 3D printing.

WRITTEN BY **ANDERSON TA**

Price as tested $799 assembled

Print volume 5"×6"×5"

Heated bed? No

Print materials PLA, nylon, Laywood

OS supported Linux, Mac, Windows

Print untethered? With SD card, initiated from computer

Open-source hardware? Yes

Open-source software? Yes

Printer control software Repetier-Host or Pronterface

Slicing software Slic3r

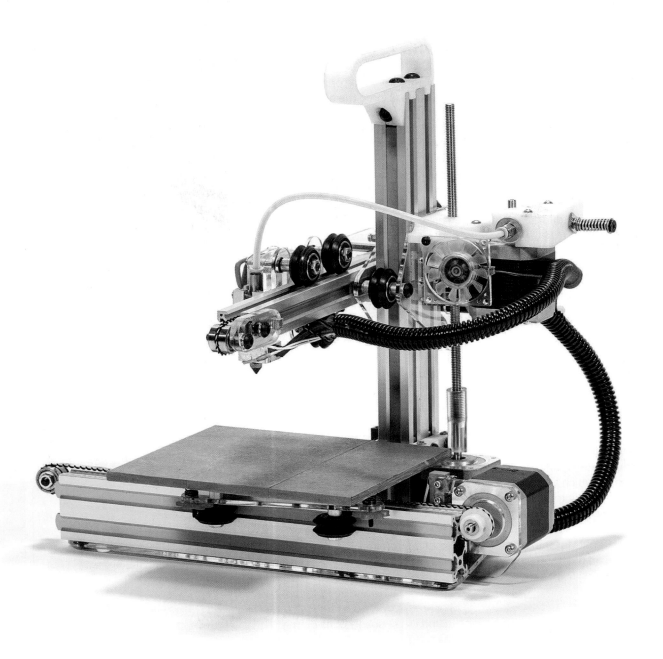

Gregory Hayes

Following up the Kickstarted success of their Bukobot 3D printer, Deezmaker, an open-source 3D printer store and hackerspace in Pasadena, Calif., has created something entirely new with the Bukito. We consider it the new standard for portable 3D printing: small, simple, and high-performing. (For more on Deezmaker, see page 106.)

Portability Without Sacrifice

With portability, there usually come sacrifices. That's not really the case with the Bukito, as it was designed and optimized for printing on the go. Its build volume of 5"×6"×5" is small when compared to its big brother the Bukobot, but it's enough for most users' printing needs. And it's rock solid, using V-slot extrusions (another Kickstarted project, by OpenBuilds) for its frame and linear motion components. V-slot is an extruded aluminum profile with chamfered interior slots for plastic wheels to ride along, so it does double duty as a structure and motion component, simplifying construction.

Looking around the machine, you can see that Deezmaker has really built the Bukito to be mobile. There's little exposed wiring: Wires for the motors, hot-end, and thermistor are nicely routed and sleeved, and the power cables are tucked away neatly, but the power button and the power and USB jacks are still easily accessible.

Users will likely notice the lack of a heated print surface (Deezmaker has plans for upgrades), so PLA is the material of choice. This decision was also likely a result of portability, as it decreases the overall power requirements. The Bukito also comes standard with SD printing capabilities, though users still must initiate print jobs from a computer.

Unique and Innovative Hardware

Users will notice a couple of differences between the Bukito and other 3D printers out there. Notably, the Bukito uses syncromesh cables instead of the typical flat-toothed timing belts. Syncromesh has a smaller overall profile than typical timing belts, which allows it to be routed inside the aluminum extrusion slots for a more streamlined integration. It also has the added benefit of being less prone to slipping and wear.

Another difference is the use of a Bowden extruder: The drive mechanism for pushing the filament is located off to the side rather than on the carriage. This allows the Bukito to achieve very fast speeds since the print head (hot-end) doesn't have to carry around all that extra weight. During testing, we were able to increase speeds by four times the default and the Bukito didn't miss a step! No spooling mechanism is included with the Bukito, so a lazy susan is recommended for your plastic management needs.

In our testing, we ran Deezmaker's recommended settings and found the Bukito to be a really formidable machine. It was fast, quiet, and accurate with minimal fuss. One area that could use improvement was retraction, i.e. when the filament is pulled back while the print head relocates. We experienced noticeable stringing when printing multiple segmented features. This is likely a result of unoptimized retraction on the Bowden setup, and presumably will be addressed when the machine goes into production.

While we were testing, the Bukito successfully raised its fundraising goal on Kickstarter. Documentation and the usual resources were unavailable, but comparisons can likely be made in looking at the Bukobot (see our review on page 84).

Conclusion

The Bukito is a printer that's good for multiple types of users: hackable enough that enthusiasts can modify to their hearts content, but simple enough that teachers can put it on students' desks. The small footprint and overall portability make it a really adaptable machine, and the PLA-only option is good for all environments, with no unpleasant odors to contend with.

The success of the Bukito will be contingent on the quality of the setup and troubleshooting documentation for the new user, but we're impressed with this machine. ◪

We found the Bukito to be a formidable machine — fast, quiet, and accurate with minimal fuss.

HOW'D IT PRINT?

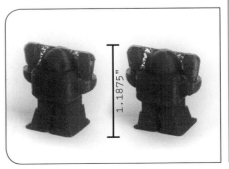

1.1875"

ULTIMAKER 2

ULTIMAKER / ULTIMAKER.COM

A bleeding-edge new machine that's great for anyone from beginner to pro.

WRITTEN BY **ERIC CHU**

Price as tested $2,565
Print volume 8.9"×8.9"×8.1"
Heated bed? Yes
Print materials PLA or ABS
OS supported Linux, Mac, Windows
Print untethered? Yes
Open-source hardware? Yes
Open-source software? Yes
Printer control software Cura
Slicing software CuraEngine

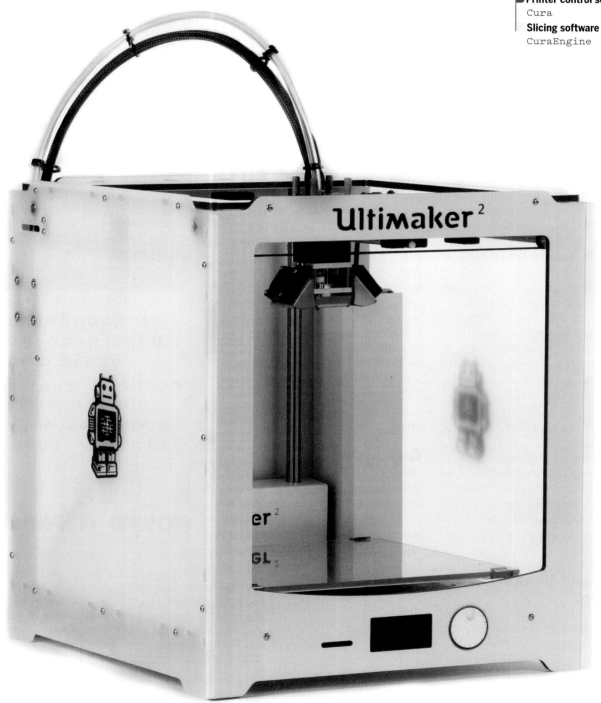

Jeffrey Braverman

Ultimaker has revamped an already excellent machine to create one of the best consumer 3D printers we've seen. The Ultimaker 2 has a bigger print volume and lots of hardware upgrades, and thanks to the company's open-source Cura software (see page 90), it's easy to use. This is a top-quality machine for people who just want to hit "Print," and it's also wide open to those who enjoy tinkering with their printer to get the most out of it.

Ultimaker was founded in May 2011 by hardcore early RepRap adopter and all-around desktop 3D printing guru Erik de Bruijn, designer Martijn Elserman, and ProtoSpace FabLab manager Siert Wijnia. Ultimaker sells kits and assembled printers, as well as add-ons and filament. In addition, they recently launched YouMagine.com, a 3D file-sharing website that features their new UltiShaper 3D modeling tool.

Looking Good!
The Ultimaker 2 is one of the best-looking printers on the market, with a frame constructed of sleek aluminum-polymer panels and frosted acrylic. LED strips illuminate the inside, and ambient mood lighting diffuses through the sidewalls when printing at night; it's quite beautiful. Simple parts and sheet-metal covers give off a clean look.

Tons of Upgrades
The new heated glass print bed supports ABS printing and is larger than before. You can print PLA directly on it without tape or other surface treatment, and after the bed cools, the printed part pops right off. (This should be standard on high-end printers.)

The redesigned extruder is now direct drive, providing quicker and quieter retraction. The hot-end is constructed of metal with a Teflon insulator; it heats up fast (220°C in 1 minute) and uses a more accurate PT-100 temperature sensor instead of a thermocouple. Dual fans cool parts more uniformly (this helped to create one of the best pairs of MAKE robot "armpits" seen during our testing).

The new electronics digitally control the motor current via firmware, rather than by a physical potentiometer. Less current is used, so the motors produce less noise.

Ultimaker's handy onboard control interface is now built into the printer and has been redesigned to be cleaner and more user-friendly, with a graphical OLED screen and a larger dial-button combo. New scripts guide the user through bed leveling and filament changing. Print settings can be adjusted on-the-fly during a print.

And the software's improved. Cura has been upgraded to slice faster than the competition, and Ultimaker has developed UltiGCode, a "flavor" of G-code that allows retraction and material settings to be changed in the printer settings (instead of in the slicer) without having to reslice.

How's It Print?
The Ultimaker 2 is even more accurate than the original. The Secret Heart Box printed very well: The hinges articulated perfectly and the outside finish looks great. Small features on the MAKE robot's back logo were sharp and crisp, among the best we printed. We did experience some ringing when printing small parts on the default High Quality setting, which set the speed too high. However, Ultimaker constantly improves their software, so we wouldn't be surprised if this issue is addressed by the time you read this review.

Keeping It Open Source
Ultimaker 2 will remain open source. From de Bruijn: "As soon as we start shipping, we will release the first major part of the system: the completely redesigned electronics (under Creative Commons Attribution-ShareAlike) ... We plan to have released everything in at most 6 months."

Conclusion
Ultimaker has delivered a bleeding-edge machine to the consumer market. The Ultimaker 2 is solid and beautiful. Its heated bed works wonderfully, and not having to apply extreme force to remove large prints is a huge advantage. Although it has a bit of catching up to do in print quality, its hardware and software upgrades make it a real contender for the prosumer crown. ◼

The Ultimaker 2's hardware and software upgrades make it a real contender for the prosumer crown.

HOW'D IT PRINT?

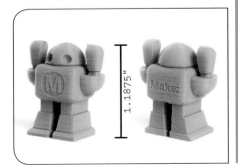

1.1875"

REPLICATOR 2

MAKERBOT / MAKERBOT.COM

Last year's impressive machine improves — with key upgrades from the open source community.

WRITTEN BY **ANDERSON TA AND BLAKE MALOOF**

Price as tested $2,199

Print volume 11.2"×6"×6.1"

Heated bed? No

Print materials PLA

OS supported Linux, Mac, Windows

Print untethered? Yes

Open-source hardware? No

Open-source software? No

Printer control software MakerWare

Slicing software MakerBot Slicer

Gregory Hayes

MakerBot Industries, based in Brooklyn, N.Y., has been in business longer than any other desktop 3D printer manufacturer, and it shows in the Replicator 2, among the most mature of the fused-filament fabrication printers in the desktop market today. Their fourth-generation printers stands out in a room full of plywood and aluminum extrusions as one of the few that looks and works like a consumer product.

Untethered and Intuitive

With its black powder-coated frame, black PVC panels, and soft glowing LEDs, the MakerBot Replicator 2 is attractive and durable. It has onboard controls with a large LCD screen that enable completely untethered printing directly from an SD card, so you can print all day long without having to attach your computer. MakerBot's MakerWare software, which is beautiful, intuitive, and easy-to-use, gets you printing quickly.

Last year we judged this machine "best in class" in the premium category — this year's model is the same great machine, with a few key hardware, software, and firmware upgrades. Interestingly, although the machine and its software are now closed source, these improvements have all come from open-source projects.

Extruder Fix Now Standard

The initial design of the Replicator 2 extruder had tensioning problems that often resulted in filament feeding failures. Through Thingiverse, MakerBot's 3D file-sharing website, an openly designed and well-publicized extruder upgrade emerged, due to the efforts of superstar Thingiverse users Dr. Henry P. Thomas (whpthomas), Emmett Lalish (emmett), and Deezmaker's Rich Cameron (whosawhatsis). MakerBot decided the approach was "too good to ignore" and began offering it as an upgrade kit for the Replicator 2. The upgraded extruder, with a spring-loaded arm that squeezes the filament between the drive gear and a bearing, is now standard on the Replicator 2.

In addition, the acrylic build plate

that was formerly covered with a raised MakerBot "M" logo has been replaced by a completely smooth plate. In the previous version, the logo would emboss itself into the bottom of prints.

Firmware Improvements

The biggest change in the Replicator 2 from last year's model is the upgraded MakerBot firmware. We noted greatly improved acceleration, better quality on high-detail prints, and an overall quieter printing experience. These improvements came directly from the Sailfish firmware project, which was created and is maintained by the dedicated efforts of two exceptional developers, "Jetty" and "Dnewman," who have openly shared their firmware (if not their identities) with the Thingiverse community.

Software Updates

MakerBot's proprietary software MakerWare has also been given some love. The MakerBot Slicer (formerly called "Miracle Grue") has been overhauled to slice faster and more accurately, for improved print quality. An "auto layout" option for automatically arranging models on the build plate has been added. Another welcome addition is that slicing profiles for both of the integrated slicers (MakerBot Slicer and Skeinforge) are now fully editable.

Conclusion

All these improvements made a big difference in print quality. The machine "just worked" without fuss throughout our testing. PLA prints had no stringing and a uniform finish with no gaps or ridges. The Replicator 2 printed the best Heart Box in the shootout, and the second-best half-size Lightbulb.

For ease of use and excellent print quality right out of the box, the MakerBot Replicator 2 is still the printer to beat in the prosumer market. MakerBot CEO Bre Pettis recently said in a MAKE Google Hangout that his machine is "good enough for a professional to put on their desk, but friendly enough for everyone else." The Replicator 2 lives up to this claim, although at $2,199 it remains out of the price range of many users. ◪

Upgraded extruder, improved acceleration, better high-detail prints, and an overall quieter printing experience.

HOW'D IT PRINT?

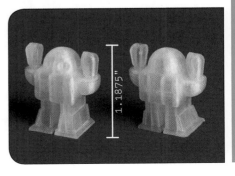

1.1875"

UP PLUS 2

DELTA MICRO FACTORY / PP3DP.COM

Automatic platform leveling takes this trusty printer to the next level.

WRITTEN BY **MATT STULTZ**

- **Price as tested** $1,649
- **Print volume** 5½"×5½"×5¼"
- **Heated bed?** Yes
- **Print materials** ABS, PLA
- **OS supported** Mac, Windows
- **Print untethered?** Yes, SD card
- **Open-source hardware?** No
- **Open-source software?** No
- **Printer control software** Up Software
- **Slicing software** Up Software

Gregory Hayes

For newcomers to 3D printing, two techniques are always a challenge: leveling the printer's build platform, and setting the proper height of the platform relative to the extruder nozzle. These procedures are usually done by hand, they're often ambiguous, and doing them wrong can result in bad prints or damage to your printer. The new Up Plus 2 changes all of this.

Awesome Auto Calibration

The Up Plus 2 comes with an all-new auto calibration procedure. Unlike other printers that simply move the print head around while asking you to slide paper underneath and turn adjustment screws, the Up uses a magnetically attached instrument to measure its own platform at multiple locations, so it can compensate in software for any tilt. A secondary sensor measures the distance from platform to nozzle. After these readings are taken and the compensations calculated, you can confidently run your prints without the worries and hassles that come with an uneven build plate.

Nice Custom Software

The Up's software has features we love, like automatic centering and placement on the print bed, and it makes it easy to add files to a print job and adjust them for printing. The new 2.0 version handles auto-leveling and improves workflow and ease of use, with new icons to guide you through the printing process (though 2.0 wasn't yet available for Macs at press time).

This great software can also be a downside. For advanced users who know the abilities of their machines and want to push them to their fullest, the closed-source software is limiting. For example, the Up automatically generates support material to prevent overhangs from drooping and ruining your print. You can alter the rules that generate it, but you can't fully turn it off. The printing of extra support structures results in an excess of material being used, as well as slower print times.

Mixed Print Quality

Print quality was very good on the models we tested, and the "raft" that's

AFINIA H-SERIES

MICROBOARDS TECHNOLOGY
AFINIA.COM

Our top pick in last year's guide was the Afinia H-Series printer, a repackaged Up Plus sold in the U.S. market, with a great U.S. company-backed 1-year warranty to go with it.

So this year we expected the Up Plus 2 to meet the standard established by the Afinia, and we weren't proven wrong. The Up's new auto bed calibration and 2.0 software took it to the next level, leapfrogging the Afinia slightly. We hope to see Afinia pick up this upgraded unit and start selling it also.

Even without these upgrades, the Afinia we tested this year is still a great machine that continues to live up to its reputation for print quality, reliability, and ease of use.

■ **Price** $1,599
□ **Who's It For** Makers, Designers, Makerspaces

printed to compensate for any tilt of the bed was easily removed by hand. Support material was sometimes difficult to remove, but left behind very little surface scarring. We did find a few small "burnt patches"— the Up runs its extruder hotter than most printers, and when bits of plastic stick to it and then rub off on the print, they can cause dark brown spots. However, fellow testers who own Up printers (or their Afinia brethren) say that they rarely see this on their printers at home.

Conclusion

If you're new to 3D printing or just want a machine that prints right out of the box without the hassle, the Up Plus 2 is a great pick for you. If you're an experienced user the software limitations might be frustrating, but you may find that the Up Plus 2 is a nice workhorse addition to your stable of printers. I'd love to add it to mine! ☑

PRIMO FEATURES
■ Innovative automatic platform leveling and height calibration. Every printer should have it!
■ Good print quality
■ Feature-rich, easy to use software

WHO'S IT FOR?
Beginners, Designers, Makers, Makerspaces

PRO TIPS
■ Auto calibration is easy to use, but the instructions are vague. Here's how:
1. Attach the leveling accessory to the extruder.
2. Detach cable from heated bed and connect it to the leveling accessory.
3. Turn the bot on and connect your computer. The printer does the rest!
■ If the ABS looks white where you remove the raft or support, warm it with a heat gun to restore color.

Whether you're new to 3D printing or just want a machine that prints without the hassle, the Up Plus 2 is for you.

HOW'D IT PRINT?

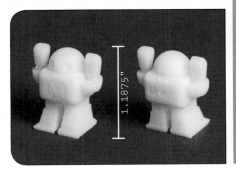

1.1875"

CUBE 2

3D SYSTEMS / CUBIFY.COM

Safe and easy to use, but a closed system holds back its potential.

WRITTEN BY **BLAKE MALOOF** and **MATT STULTZ**

- **Price as tested** $1,299
- **Print volume** 5½"×5½"×5½"
- **Heated bed?** No
- **Print materials** PLA, ABS
- **OS supported** Mac, Windows
- **Print untethered?** Yes, over wi-fi or via USB flash drive
- **Open-source hardware?** No
- **Open-source software?** No
- **Printer control software** Cube software
- **Slicing software** Cube software

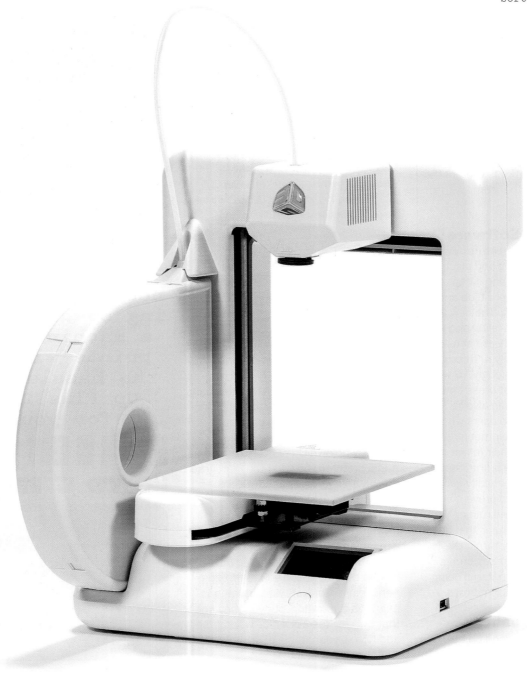

Gregory Hayes

Last year we reviewed 3D Systems' entry into the desktop 3D printer market, the Cube. It was a relatively affordable, compact, simple-to-use printer that used closed-source software and proprietary filament cartridges. These elements made this printer less appealing to 3D printing veterans, but great for young or inexperienced makers who want to venture into the world of 3D printing.

This year 3D Systems released the Cube 2, adding a few modifications that make it even safer, quieter, and more kid-friendly, in hopes of better meeting the needs of the education market. Unfortunately the price paid for these changes is reduced print quality.

Very Easy Setup, Nice Touchscreen Features

The Cube 2 is very easy to set up. The required software can be downloaded from cubify.com and supports both Mac and Windows (sorry, Linux users). After installation, a simple graphical interface walks you through the steps to prepare a 3D model for printing. The software also includes a "heal" option that will fix flaws in the selected model to make it print-ready.

The software lets you make some basic changes, such as scaling and orientation of the model, before exporting the model to the provided thumb drive, which can be used to print without the need for an attached computer. The Cube also includes a built-in wireless network that allows models to be sent to the printer over wi-fi.

The built-in touchscreen interface allows you to quickly access useful features like bed leveling, nozzle calibration, and filament loading/unloading. (The Cube uses proprietary cartridges, making its filament more expensive than other printers.) With the thumb drive plugged into the printer, you can select the model you wish to print and start the process from the touchscreen.

Quieter and Safer Than Last Year, but Print Quality Suffers

The first thing we noticed when the printer fired up was the motors used in this version of the Cube are sig-

nificantly quieter than those in last year's model. This version also sports a silicone bumper around the extruder nozzle, which helps protect any curious fingers from being burned by the hot-end. It also keeps stray filament from coiling up and sticking to the nozzle, a common problem in other printers.

3D Systems has included a magnetically attached glass build platform; the lack of a heated build platform adds one more level of safety, but means the Cube needs a little more help getting prints to stick to the build plate. This help comes in the form of CubeStick glue, a solution that's applied to the glass plate with the included squeegee bottle. This is easily the most tenacious grip we've seen of any print to a platform. Once the print is complete, the glue dissolves in water, allowing the print to pop right off the build plate surface.

The finished print quality, however, has considerable downsides when compared to printers in the same price range. Overhangs and caps showed significant drooping, small details were lost, and the prints frequently feature a small pool of plastic on the edge of the model where the extruder started the print and placed an anchor within the print itself.

Most of the issues that we experienced were caused by the slicing software and not by hardware, meaning that with a little bit of work and a software update, this machine could be cranking out the kind of prints we would expect from it. Unfortunately since the software is closed source, users won't be able to make these changes themselves.

Conclusion

3D Systems has taken a lot of care to design a printer that can be used in educational settings without the usual safety concerns that come with most 3D printers. Parents looking to purchase a printer for a young maker will be happy with the simplicity of use in addition to these safety features. More advanced users will quickly find that the closed nature of the Cube 2 and its disappointing print quality makes this not the printer for them. ◢

With a little bit of work and a software update, this machine could be cranking out the kind of prints we would expect from it.

HOW'D IT PRINT?

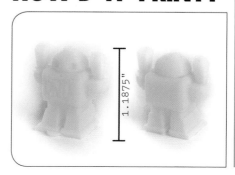

1.1875"

UP MINI

DELTA MICRO FACTORY / PP3DP.COM

A heated bed and a sleek enclosure offer an affordable multimaterial option.

WRITTEN BY **JAMES CHRISTIANSON** *and* **TOM BURTONWOOD**

Price as tested $899

Print volume
4.7"×4.7"×4.7"

Heated bed? Yes

Print materials ABS, PLA

OS supported Mac, Windows

Print untethered? Yes, after file is sent from computer

Open-source hardware? No

Open-source software? No

Printer control software
Up software

Slicing software
Up software

Gregory Hayes

Looking like a futuristic kitchen appliance, the $899 Up Mini from Chinese company Delta Micro Factory (PP3DP) is no DIY kit project — it's designed entirely with the average consumer in mind. It can print compact volumes up to 4.7"×4.7"×4.7" on a removable perf board, which sits on a heated PCB bed.

Pushbutton-Simple Setup and Software

Setting up this printer is almost as easy as plugging in a microwave and pressing the popcorn button. Just connect the printer and install the Up software from the website; it's easy to navigate and download the latest version for Mac or PC. Grab the PDF manual while you're there.

The leveling software does a great job of getting the nozzle and build platform aligned, but be ready to shim the platform as there is a tendency for it to dip at the front left just a touch.

Enclosed for Safety, Cozy for ABS Printing

Up Mini's enclosed housing helps to reduce the risk of kids (or clumsy people) coming into contact with the heated components during operation. Retaining the heat also helps prevent ABS prints from warping. The enclosure makes the build area a bit smaller, but on the bright side, there's less worry of software-generated rafts extending off the bed since the nozzle can't quite travel to the edges of the platform.

Although the printer is enclosed, you can still watch it print. There are two hinged doors, the top one providing a better view during operation. That door also provides filament access — the plastic runs off a spool mounted on an arm in back, then passes over the top of the machine, down through the joint of the top door, and into the print head.

Mostly Quiet Operation

Aside from the obnoxiously loud buzzer that sounds at initialization, print start, and print end, the Up Mini's housing allows for very quiet operation. The button on the front with blinking indicator light, and the LED work light inside, are well-designed features. The printer

operation is similar to its big brother, the Up Plus 2 (see page 70), with the addition of a magnetically removable print head. Although there are no onboard controls, once the print starts you can disconnect the computer.

Extruder Jams

Once the file is sent and the print begins, you can let the Up Mini run unsupervised and you'll rarely see an "air print" — as long as you're using their filament. However, over the course of our testing weekend, we experienced at least three filament jams that we weren't able to diagnose conclusively. It's possible the problem was with our particular spool of filament, how the filament was being fed into the head, or heat building up inside the housing.

We also had some adhesion issues between layers on our smaller prints (the MAKE robot). Experienced Up Mini users informed us that this was unusual. They also advised that the filament guide at the back of the machine should be avoided and that it's helpful to print with the top door open, to vent excessive heat in the enclosure.

Solid Objects Printed Best

We printed in ABS; PLA is also possible but experienced users advise adding a fan to cool it. On our first print we had some difficulty removing supports around delicate extrusions and keeping parts intact. However, the Up Mini handled detailed and thin-walled structures well. The calibration cube turned out sharp when using the Fine software setting, and other solid objects printed well, but the Spiral Lightbulb torture print was something of an overhang mess on its last section. We tried it a couple times, always with similar results.

On some occasions, not being able to fully turn off the raft caused unnecessary cleanup. However, the rafts come off cleanly, and the precision of supports is very well calculated.

Conclusion

The Up Mini is an easy-to-use, affordable, complete package. Having the ability to print in ABS gives this little guy the upper hand in its price range. ◪

The ability to print in ABS gives this little guy the upper hand in its price range.

HOW'D IT PRINT?

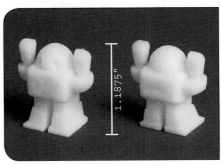

1.1875"

FELIX 2.0

FELIX ROBOTICS / SHOP.FELIXPRINTERS.COM

Smart design, quality construction, and immaculate prints out-of-the-box.

WRITTEN BY **KACIE HULTGREN AND MATT STULTZ**

Price as tested $1,949 assembled

Print volume 10"×8"×9¼"

Heated bed? Yes

Print materials PLA, ABS, nylon

OS supported Linux, Mac, Windows

Print untethered? Yes, onboard controls

Open-source hardware? No

Open-source software? Yes

Printer control software Repetier-Host

Slicing software Skeinforge / SFACT

Gregory Hayes

A sleeper hit in our testing, the second-generation 3D printer from Netherlands-based Felix Robotics delivers on the promise it showed last year, with impressive hardware upgrades and greatly improved performance. Its extruded aluminum frame, augmented by 3D-printed parts, provides a simple and open design, and its heated aluminum build plate can hold large builds.

Easy Setup

The downloaded PDF instruction manual covers assembly, calibration, and software, with clear photos and step-by-step instructions. Felix works with the typical open-source controllers Repetier-Host and Pronterface, and they provide a package of Repetier-Host conveniently paired with SFACT slicing profiles. Setup was quick and easy in Windows; for Mac users, Felix recently added a Repetier download and instructions for installing SFACT profiles separately, though we weren't able to test these.

Great Prints Right Out of the Box

Our out-of-the-box prints were immaculate. The Felix's print quality shined on the small, overhanging diameters of the half-sized light bulb. The tolerances of our preassembled Heart Box were spot-on — straight off the platform, the box opened and rotated without any clean-up. SFACT uses a modified Skeinforge slicing engine that's slower than other slicers, especially on large or complex prints. The provided profiles, and the machine itself, are optimized for 1.75mm PLA, but its heated bed and high-temp extruder (rated to 275°C) mean you can print ABS and other compatible plastics without hardware upgrades.

Felix ships with a 0.35mm nozzle, slightly smaller than the standard 0.4mm found on many machines — this may be partially responsible for the successful delicate prints, but it also adds increased print time for solid fill layers. Felix Robotics runs their production printers at 80mm/s and recommends 30–40mm/s for high-quality prints. At those speeds, it wasn't the fastest machine in our shootout, but it felt quick and ran quiet, and a pro user could push it faster. The company lists a maximum speed of 200mm/s.

The large, heated aluminum print bed needed patient adjusting, but the built-in three-point leveling system was up to the task. There's no spool holder, so we used the supplied spinner, which was ingeniously simple but needed frequent attention to avoid kinks and misfeeds. The aluminum frame and open design provide great strength and rigidity without needing a lot of extra construction, and a built-in handle makes transport easy. The Felix 2.0 has also upgraded all of its linear motion systems to top-of-the-line ball bearing slides that keep things perfectly aligned and smooth.

Onboard Controls and Upgrades

A mini SD card and integrated LCD display come as an optional upgrade with the Felix, for untethered printing. As of press time, there was no documentation available for the LCD display, but the operation is fairly straightforward.

It's worth noting that Felix Robotics sells kits for converting last year's Felix 1.0 and 1.5 to a 2.0. Kits are customizable so the user can choose which improvements are most important: printed parts, new motors, a heated bed, LCD, and more. If future upgrades are important to you, it's nice to know this company is providing its early adopters with a path.

New Model: Felix 3.0

A new Felix, version 3.0, was not yet available for our review, but the company promises dual extrusion, a flatter heated bed, a new hot-end with exchangeable brass tips for easier cleaning, and more.

Conclusion

The Felix 2.0 has lot to be proud of — the prints were among the best in our shootout. These great results come with time tradeoffs due to the slicing profiles, but the hardware had few hangups, and the LCD upgrade promises untethered prints, important for printing big objects without a dedicated computer. If you're looking for an upgradeable machine and a company that will make sure it can happen, the Felix 2.0 is an excellent printer for you. ◪

The Felix 2.0 is a sleeper hit with a lot to be proud of. Its prints were among the best in our shootout.

HOW'D IT PRINT?

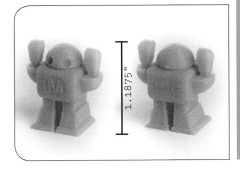

1.1875"

TINKERINE STUDIOS
DITTO+

TINKERINE STUDIOS / TINKERINES.COM

Big and capable, with high-quality prints.

WRITTEN BY **KACIE HULTGREN**

Price as tested $1,549 assembled ($1,249 kit)

Print volume 8¼"×7¼"×9"

Heated bed? No

Print materials PLA

OS supported Mac, Windows

Print untethered? Yes, SD card and onboard controls

Open-source hardware? No

Open-source software? Yes

Printer control software Coordia

Slicing software Skeinforge integrated into Coordia

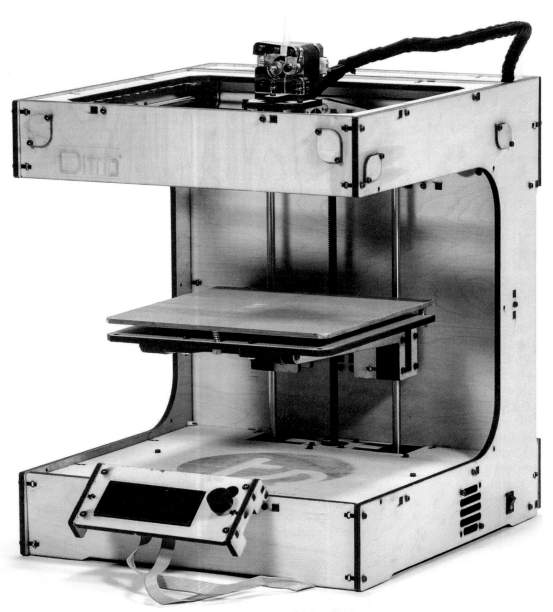

Gregory Hayes

The wood-framed Ditto+ is the flagship printer for Tinkerine Studios of Vancouver, B.C. Their first model was launched through Indiegogo in the spring of 2012, and the Ditto+ marks their second-generation upgrade.

The Ditto+ can accommodate large prints up to 8¼"×7¼"×9" at resolution ranging from 100–300 microns. We did most of our test prints with their medium resolution profile.

Hardware Details Could Use Some Attention

The C-shaped frame provides an open design that allows a great view of your print in progress, though it's somewhat obstructed at the start. The print bed has a three-point leveling system with springs, which works well, but we found the adjustment nuts rough on our fingers — they cry out for a printed thumbwheel upgrade.

The Ditto+ has a built-in spool holder on the back that should fit many spool sizes and includes a feed tube to the extruder. We found the filament needed a little extra encouragement to feed into the extruder. On more than one occasion, the rod fell out of the back slider on the x-axis; we fixed it mid-print without pausing and it didn't seem to impact print quality. It's worth noting that the extruder takes longer to heat up than other similar models.

Onboard Controls

We appreciate the inclusion of an integrated SD card slot — with a large build area, it's nice to be untethered. The functions of the LCD display aren't documented in the online manuals, but most of the features are straightforward. You can preheat, start a print, monitor the print, jog the motors, and more from the interface alone. We did find the jog mode a little fussy — the z-axis and extruder jogs were easy to activate at the same time by accident, and the y-axis on occasion slowed to a crawl.

The two-part online PDF manual covers assembly and calibration with straightforward line drawings and step-by-step instructions. Though we tested a prebuilt unit, the assembly and calibration guides are complete and well documented. Step-by-step instructions lead you through your first print, and a troubleshooting guide covers most common problems and solutions for rectifying them. There's also a well-written explanation of slicing software settings and how to change them, to get new users up to speed quickly.

Some Software Limitations

Tinkerine Studio's free, downloadable open-source software Coordia (still in beta) is largely uncomplicated, using Skeinforge for slicing. The addition of PyPy speeds this process and, conveniently, it installs automatically. The software also contains a built-in 3D G-code visualizer.

Coordia has a few limitations, however. It doesn't have a plating function or placement tools. Clicking on Slice will open, place, and slice your file in one fell swoop. This provides simplicity for a new user, but might be frustrating for a user who wants more control. If you'd rather use more common slicing and control software, Tinkerine Studio recommends Pronterface and Slic3r, but at press time, there are no supplied Slic3r profiles.

We encountered an extrusion problem that was solved with a quickly answered phone call to technical support. They deftly diagnosed a software bug that was stopping the extruder motor mid-print. With a couple of altered slicing settings, we solved our extrusion issue and saw a vast improvement in print quality.

Conclusion

If the prosumer machines are out of your price range, the Ditto+ may be the ticket. The machine can print at accelerated speeds, the build area is big enough for most projects, the documentation and software are simple enough for entry-level users, and the integrated display and SD card interface give it versatility. And although prints weren't entirely without problems, the Ditto+ print quality was a cut above, performing as well or better than similarly priced printers with out-of-the-box settings. ◪

If the prosumer machines are out of your price range, the Ditto+ may be the ticket.

HOW'D IT PRINT?

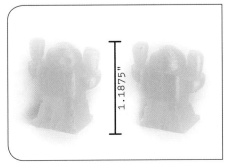

1.1875"

TYPE A MACHINES
SERIES 1

TYPE A MACHINES / TYPEAMACHINES.COM

With bold design and progressive wireless printing, it could be a 2014 frontrunner.

WRITTEN BY **MATT GRIFFIN** *and* **CHRIS McCOY**

Price as tested $2,295 (expected, 2014 model still in beta)

Print volume 12"×12"×12"

Heated bed? No

Print materials PLA, nylon, soft PLA, PET

OS supported Linux, Mac, Windows

Print untethered? Yes, wi-fi and Octoprint

Onboard control interface No

Open-source hardware? Some restrictions

Open-source software? Yes

Printer control software Octoprint

Slicing software KISSlicer

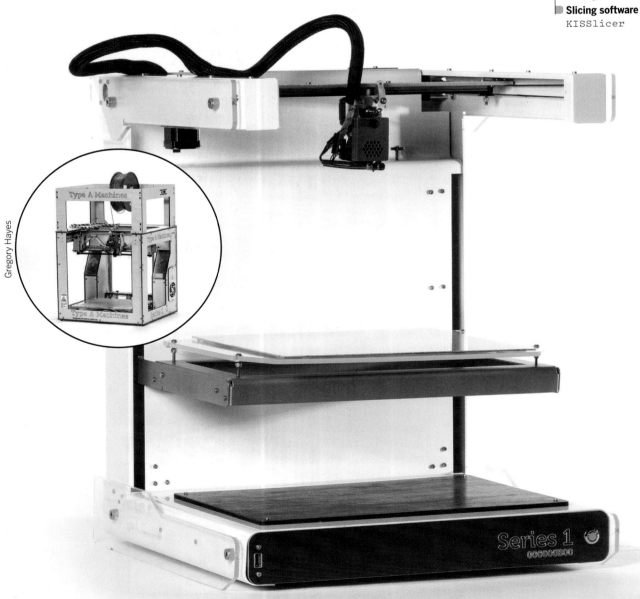

Gregory Hayes

Jeffrey Braverman

Type A Machines was born out of the famous San Francisco hackerspace Noisebridge. Andrew Rutter started the company in December 2011 (see page 25) and launched the Series 1 printer the following spring. Type A describes its philosophy as "accessible source": They will hold patents, but make all hardware specifications fully and publicly available for user modification (with commercial restrictions).

Most of our tests ran on Type A's 2013 model Series 1. A new 2014 model is expected in October, and the 2013 model will no longer be available; we also tested a beta prototype of the 2014 machine, but had only a few hours with it. This review is split into two sections describing both models.

Out with the Old: 2013
The 2013 Series 1 ($1,695) is a significant upgrade from the 2012 model we rated "Best in Class" last year. It includes redesigned mechanics, an improved version of Type A's Winchester extruder, a custom machined pinchwheel, better motors, and a four-point build platform leveling system. The 2013 Series 1 also ships with a handy Top Hat spool-management system, so you can queue up a rainbow of spools and take advantage of the Winchester's filament quick-change lever.

Our unit's bed was out-of-level on delivery, so we releveled and recalibrated following the manual. We had little or no adhesion issues with this printer for the whole weekend, which was unusual among those we tested. The factory-supplied KISSlicer profiles worked well with very little adjustment. A slider controls the print speed, from Fast (90mm/s) to Precise (30mm/s). We found that pushing the speed past 65mm/s caused registration and surface-finish problems.

We used only one spool, so we didn't test the Top Hat spindle for swapping. The spool occasionally unwound during homing, but it didn't cause issues during printing. We tested using the bundled 1.75mm red PLA. After our configuration-sequence print, we went immediately for the Ultra Fine 50-micron layer height slicing profile and saw some problems with underextrusion,

strings, and retraction. Type A support advised tweaking the slicing settings, lowering the speed and temperature, and measuring the filament's real width using calipers. This advice was spot-on; our prints looked great afterwards, even on Ultra Fine.

In with the New: 2014
Compared to the 2013 model, the 2014 Series 1 ($2,295) boasts a clean industrial design, huge volume (the biggest in our tests), enhanced gantry mechanics, a glass build platform, and a new sheet-vinyl surface overlay that might just dethrone blue painter's tape for printing PLA. A particularly interesting new feature, developed from user feedback, is an easy-to-remove, low-cost extruder nozzle cheap enough to be considered "disposable" if it clogs. Type A also promises a new "hacker-friendly" warranty that they'll honor even if you've modified the machine.

Though the beta-stage prototype we tested did not have an onboard display (one was being considered for production), it did have another innovation — built-in wi-fi control via Octoprint, an open-source, web-based printer client that sends G-code and controls the printer via an embedded Linux platform. Installing Octoprint typically requires use of the Linux command line, but the Series 1 does it for you — it's got a BeagleBone Black single-board computer with Octoprint preinstalled.

Our single test print showed minor PLA retraction issues that resulted in undesired stringing between free parts of the model, but we had only a few hours with a beta machine, so we can't yet judge print quality. It's worth reiterating that Type A's 2012 and 2013 machines printed very well.

Conclusion
The Type A Machines 2014 Series 1 sets itself apart with the addition of integrated wireless printing, a sleek industrial design, and hacker-friendly warranty. It's also slated to be compatible with direct printing from Windows 8.1. We look forward to this machine's official release, anticipated in late October; check makezine.com/review for an updated review. ◪

Type A Series 1 is the first printer to include an embedded Linux wi-fi board with Octoprint preinstalled, for open-source wireless printing right out of the box.

HOW'D IT PRINT?

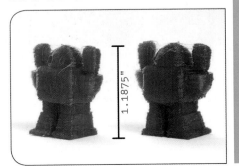

1.1875"

LULZBOT TAZ

ALEPH OBJECTS / LULZBOT.COM

A powerhouse for experienced users, with huge, fast prints for those willing to tinker.

WRITTEN BY **JOHN ABELLA**

- **Price as tested** $2,195
- **Print volume** 11.7"×10.8"×9.8"
- **Heated bed?** Yes
- **Print materials** ABS, PLA, PVA, HIPS, and Laywood
- **OS supported** Linux, Mac, Windows
- **Print untethered?** No (1.0), Yes (2.0)
- **Open-source hardware?** Yes
- **Open-source software?** Yes
- **Printer control software** Printrun
- **Slicing software** Slic3r with SFACT profiles

Gregory Hayes

Two questions I always hear when demonstrating 3D printers to the public are: "Can it print faster?" and "Can it print bigger?" The LulzBot TAZ feels like it was created by people who wanted their answer to be emphatic.

Solid Hardware, Huge Build Volume

The TAZ print area is 11.7"×10.8"×9.8" for a whopping build volume of 1,238 cubic inches, the second biggest of any machine we tested this year. You could print a life-sized basketball with room to spare. The footprint is substantial — it'll fit on your desk, but not much else will.

The TAZ also had the most 3D-printed parts of all of the machines we tested — brackets, knobs, enclosures, even two-color printed company logos. (LulzBot maintains an army of printers to keep these parts in production — so they definitely know what it takes to print working parts in volume.)

The frame is black anodized aluminum extrusion with 3D-printed connectors; the mechanics are exposed. Cabling and electronics are well managed; there are no loose wires or exposed connections, and the main electronics are in a vented, 3D-printed case.

Exceptional Documentation

One of the overriding themes in last year's *Ultimate Guide to 3D Printing* was that 3D printers needed better documentation, and LulzBot has come back with a vengeance. The TAZ manual is not only the most thorough, nicely bound, and well-designed in the bunch, it's also a great primer to 3D printing, path generation, tips and tricks, and tuning. This book would be useful to anyone using a 3D printer, and I hope they offer it for sale; I'd buy one.

Similarly, LulzBot included the best set of tools — everything you need to tweak, tune, tighten, remove prints, and do regular maintenance. This is just attention to detail at an epic level.

Some Printing Snags

Our test unit didn't print great with the stock profiles from the website, and required tweaking before it was printing as we expected, but LulzBot support was helpful in ironing out our toolchain and configurations. We tested the unit with 3mm ABS filament provided by the vendor; it should print PLA just as well.

We noticed the print bed was significantly hotter in the center than at the outer areas. This temperature gradient didn't cause problems for us but it needs to be watched with larger ABS prints, as it could cause issues with first layer adhesion.

Fast and Quiet

The TAZ printed quickly and quietly; stock settings were faster than average, and it'll go even faster if you're willing to do some tuning. The software toolchain was the common RepRap-centric open source combination of Slic3r and Printrun.

The TAZ also has a very nice integrated spool holder and filament management system with adjustable tension. Our prints ran reliably, with no failed jobs or extruder jams.

The TAZ 1.0 doesn't have an SD card slot, so all our prints needed to be done over USB. With a build area this large, some prints could go out past the 30-hour mark; for a user with a laptop, that would mean leaving it tethered the whole time. We're glad to report LulzBot is remedying this with their upcoming 2.0 model.

New Model: TAZ 2

Since our testing, Aleph Objects has upgraded to a TAZ 2 with an improved hot end and a new LCD screen and controller with an SD card slot for untethered printing. It's slated to go on sale in mid-October; look for our review online at makezine.com/review.

Conclusion

LulzBot goes out of its way to support open hardware and software initiatives, even helping to fund development of Slic3r, the path planning software used by many printers. Plans for building a TAZ are freely available, and scratch-built versions are showing up online from enterprising hackers. However, we suspect you get more for your money by buying the official TAZ. The LulzBot total package is greater than the sum of its hardware parts. ◪

The TAZ printed quickly and quietly; stock settings were faster than average, and it'll go even faster if you do some tuning.

HOW'D IT PRINT?

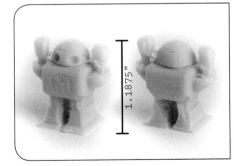

1.1875"

◾ Sneak peek: LulzBot TAZ 2

BUKOBOT 8v2

DEEZMAKER / DEEZMAKER.COM

Innovative, accurate, and fast, here's a printer for the technology lover.

WRITTEN BY **JOHN ABELLA**

- **Price as tested** $1,549 assembled ($1,299 kit)
- **Print volume** 8"×8"×8"
- **Heated bed?** Yes
- **Print materials** ABS, PLA, nylon, polycarbonate, PVA, HIPS, Laywood, Laybrick
- **OS supported** Linux, Mac, Windows
- **Print untethered?** With SD card, initiated from computer
- **Open-source hardware?** Yes
- **Open-source software?** Yes
- **Printer control software** Repetier-Host
- **Slicing software** Slic3r

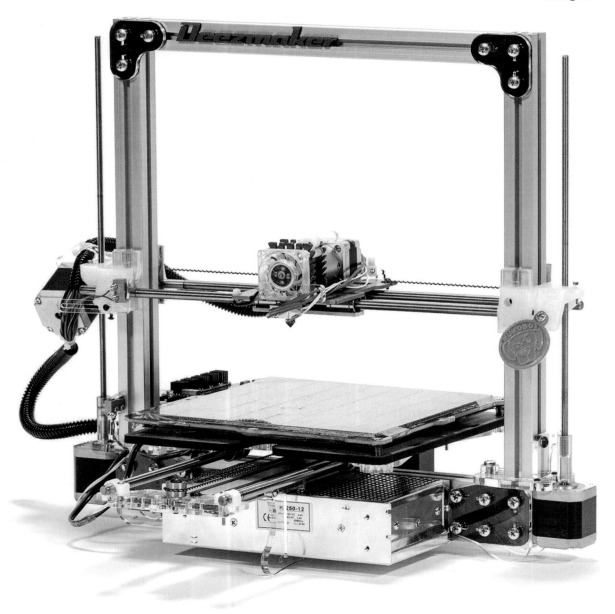

Gregory Hayes

Named after the company founder's dog, the Bukobot returns for the second year to the MAKE Ultimate Guide to 3D Printing. Was it well behaved, or did it come back to bite us? We spent significant time with the Bukobot to learn what makes it tick.

Updated and Upgraded

An updated version of last year's Bukobot, the machine is back with the same 8"×8"×8" build area, for a usable volume of 512 cubic inches. The frame, made almost exclusively of aluminum extrusions, now has fewer printed parts, some of them swapped out for laser-cut acrylic.

The footprint is about average and portability is decent; we were able to move the machine around without impacting calibration. Crucial parts are exposed and could get snagged in transit, so using the included semi-rigid travel case is your best bet. Notably, the Bukobot and its smaller sibling Bukito (see page 64) were the only machines in the shootout to come with travel cases for portability.

Unique Hardware Design

There's no getting around the look of the unit — and our testers were black and white in their take on it. "Future industrial," said some. "The work of a mad scientist," said others. The electronics and power supply are exposed, as is much of the wiring — and while it's all well handled, it does give the machine a sci-fi look.

Despite its polarizing appearance, the Bukobot got glowing reviews for its mechanical design considerations, many of which were unique among printers we tested. Deezmaker uses synchromesh cable for transferring motion instead of the belts used by every other vendor. The cables work quietly, don't wear out like belts do, and are veritably skip-proof. We also liked the z-axis sliders that fit inside the aluminum extrusions — though the long-term performance and durability of these innovative methods aren't yet well understood.

The Bukobot's documentation is wiki-style, available at bukobot.com, and seems to be current. The site has prebuilt configuration files for Slic3r and Repetier-Host, two common open-source software tools, but it doesn't go in-depth on how to operate these tools. The configurations have settings for different layer heights, infill percentages, and recommended plastics. The profiles aren't without issues, though; one that we downloaded for 3mm plastic had a filament diameter of 1.70 set in it, a misstep that could trip up an inexperienced operator.

We called tech support on a weekend and they answered questions quickly. As well, Deezmaker has a retail presence in Pasadena, Calif. (see page 106), and hosts regular 3D printing meetups.

Put a Fan On It

We tested the Bukobot with both ABS and PLA, both of which worked well. The machine prints in 3mm filament exclusively, and it can also print in both polycarbonate and nylon according to the enthusiast community. We wished it had a cooling fan to blow on the prints while using PLA — just placing a desk fan next to the unit made huge improvements in print quality.

Unlike the smaller Bukito, our Bukobot did not come with an SD card slot, meaning we had to print via USB. An SD card upgrade is available and is probably a good investment if you're going to do longer prints.

Default print speeds were above average for the group, and we pushed them significantly faster with almost no impact to print quality. The machine lacks an integrated filament spool; the company recommends a specific lazy-susan turntable, which worked well in our testing.

This isn't a printer for a first-timer, but it could be great in a makerspace or a tech-focused user group. It's not for someone with small children, either — exposed cables and moving parts are easy to get your hands into.

Conclusion

Deezmaker has been consistently out in front, getting new ideas into new printers for sale — and the Bukobot 8 v2 is no exception. The exposed wiring and electronics aren't for everyone, but if you're looking for something fast, quiet, and a little bit quirky, this is a machine to pursue. ◪

The Bukobot 8 got glowing reviews for mechanical design considerations that are unique among printers we tested.

HOW'D IT PRINT?

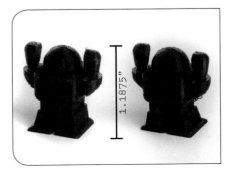

1.1875"

PRINTRBOT PLUS

PRINTRBOT / PRINTRBOT.COM

Large print volume for less than a grand equals a big value.

WRITTEN BY **NICK PARKS**

Price as tested $999 assembled

Print volume 8"×8"×8"

Heated bed? Yes

Print materials ABS and PLA

OS supported Linux, Mac, Windows

Print untethered? With SD card, initiated from computer

Open-source hardware? Yes, noncommercial

Open-source software? Yes

Printer control software Repetier-Host

Slicing software Slic3r with SFACT profiles

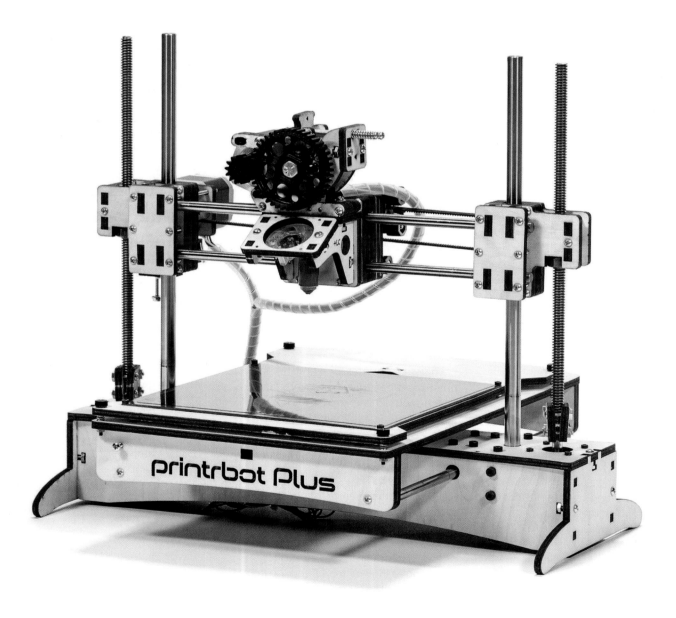

Gregory Hayes

Lincoln, Calif.-based Printrbot, founded by Brook Drumm in 2011, operates with the goal of making 3D printers affordable for every home and school (see *"Generation 3D,"* page 30). The Printrbot Plus, the grandest of the company's three machines, is a great printer for the value-minded maker — it's fast, has a large build platform, and only costs $999.

It's also a little tricky to get dialed in, so if you're looking for something that doesn't require any tweaking out of the box, this printer is probably not for you. But once the settings are fine-tuned, the Plus prints similarly to machines twice its price.

Upgradeable

One of the Printrbot Plus' attractive features is its upgradeability, great for people who aren't sure if they want to spend money on options like multiple extruders or a higher-quality bed. Having the ability to update the printer also future-proofs it by allowing new hardware to be added as it gets developed.

Support Your Objects, Level the Bed

In testing, the printer seemed to struggle on very precise parts without any support, but by using support structures and by reducing overhangs, we were able to get our prints accurate to about 0.15mm.

Using PLA gave consistently better results than ABS, which had the unfortunate tendency to peel off the bed. We had our best results with the Printrbot Plus by leveling the bed, heating it to at least 80°C and the extruder to 240°C, and encouraging print adhesion by putting a thin layer of ABS goo on the bed. (ABS goo is easily made by mixing small bits of ABS with acetone.)

Easily Correctable Over-Extrusion Issues

We noticed some overextrusion, even with the motor movement settings (aka EEPROM settings) properly calibrated. These settings make the printer move a specific distance when defined, and will change if gearing is modified or a motor is replaced. On a Windows computer, these settings can be found by clicking Config, then clicking Firmware EEPROM Configuration. We found that adjusting the filament setting to 0.1mm greater than its actual diameter eliminated these problems. Once the printer's settings were configured properly, it worked well.

And what's not to love about the 8"×8"×8" maximum build volume, larger than that of many other desktop 3D printers?

Untethered Printing Possible

The Printrbot Plus is able to print untethered, although it's not a completely straightforward process (see below). The built-in SD card reader can store files and allow users to unplug their laptops after a print has been selected. These can later be reconnected to regain control of the printer; you can then pause, stop, or modify the print. This feature works well for longer prints.

Conclusion

Overall, the Printrbot Plus is great for experienced makers and people who enjoy a lot of tinkering — it takes some fine-tuning but gets good results. And at just $999, there's a huge value here. We recommend the Plus to anyone with spare time to spend calibrating the settings, and who wants a machine with a large build volume that can support multiple materials. ◪

Once the printer settings were properly configured, it worked well. And what's not to love about a build volume larger than that of many other desktop 3D printers?

HOW'D IT PRINT?

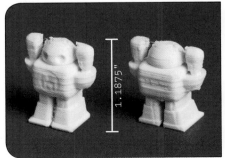

PRINTING UNTETHERED: THE MISSING CHAPTER

While the Repetier-Host software has built-in SD card support, it's slow and buggy; don't use it to transfer your G-code. Instead, unchain your computer with the following steps.

■ Use your laptop's SD slot or a card reader to copy the G-code directly from the computer to the card.
▫ Install the SD card into the printer, and connect the laptop.
■ Select the file to print in Repetier from the SD card manager, and initiate printing.
■ Click Disconnect in Repetier and unplug the USB cable. The printer will continue to run.
■ To resume control of your bot, simply plug the laptop in with USB and click Connect.

3DPRINTER 4U BUILDER

CODE-P WEST BV / 3DPRINTER4U.NL/EN/

WRITTEN BY ERIC WEINHOFFER

- ■ **Price as tested** $1,688
- **Print volume** 8.6"×8.25"×6.9"
- ■ **Heated bed?** No
- ■ **Print materials** PLA
- ■ **OS supported** Linux, Mac, Windows
- **Printer control software** Pronterface
- **Slicing software** Slic3r
- ■ **Print untethered?** No
- ■ **Open-source hardware?** No
- **Open-source software?** Yes

The Builder is a solid printer from the Netherlands. At roughly $1,688, it isn't exactly cheap, but you get a lot for your money.

The machine's print bed is quite large, 8.6"×8.25"×6.9". It can print layers as fine as 100 microns, and does so with impressive speed. The extruder is solid as well; we didn't experience any problems with it throughout the entire testing period.

One caveat: The bed sits directly on four adjustable bolts, making for easy leveling, but can cause damage to the nozzle if you bring it too high — this is where having a spring-loaded bed would be beneficial.

Aesthetically, this thing is gorgeous. The powder-coated steel frame protects the important bits and gives the Builder a polished look that you don't often find among machines in this price range.

The documentation provided is comprehensive. The PDF manual walks you through unboxing, software/driver installation, and preparing a model for your first print. And the company includes sample G-code as well as Slic3r settings, although we were disappointed to find that they included screenshots of the settings instead of a simple INI profile for importing directly into the software.

PRIMO FEATURES
- ■ Fast print speeds
- Large build area
- ■ Metal frame for sturdiness
- ■ Easy-to-remove build platform

PRO TIPS
- ■ Use simple blue painter's tape on the bed, as PLA adheres to it better than the supplied brown tape.

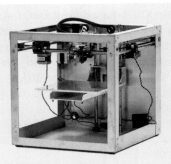

SOLIDOODLE 3
SOLIDOODLE / SOLIDOODLE.COM

- ■ **Price as tested** $799
- **Print volume** 8"×8"×8"
- ■ **Heated bed?** Yes
- ■ **Printer control software** Repetier-Host

This printer's low price is attractive, but otherwise it did not impress. The construction seems clap-trap and shoddy. Z-rods are held in place with hose clamps. The y-axis pulleys mount on a flimsy part of the sheet metal housing, causing flexure, a slack belt, and an uneasy vibration that gave problems with print adhesion. Sometimes the extruder would skip off-track with a loud stutter and ruin a print. We did see prompt responses from Solidoodle customer support.

—Chris McCoy and Eric Weinhoffer

Gregory Hayes

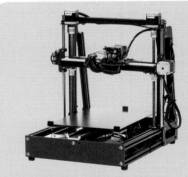

MENDELMAX 2.0
MAXBOTS / MENDELMAX.COM

- ■ **Price as tested** $2,195
- **Print volume** 7¾"×12¼"×8¾"
- ■ **Heated bed?** Yes
- ■ **Printer control software** Printrun

Sporting a sturdy aluminum frame and heated glass bed, the MendelMax 2.0 has decent print quality in the x-y plane, but the z-axis had problems like gaps and ridging. The fan is underpowered and the shroud is a weak fused-filament part. Leveling the bed requires reaching underneath with a hex key and can be tricky. The recommended speed settings seemed high; we saw quite a bit of vibration using them. Documentation was good, but this bot needs a lot of tweaking to print well.

—Eric Chu

AIRWOLF AW3D XL

AIRWOLF / AIRWOLF3D.COM

WRITTEN BY **TOM BURTONWOOD** *and* **ERIC CHU**

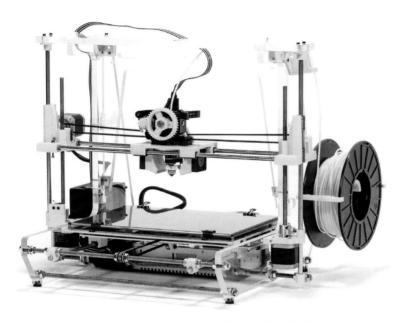

- ■ **Price as tested** `$2,399 assembled`
- **Print volume** `12"×7.9"×7"`
- ■ **Heated bed?** `Yes`
- ■ **Print materials** `ABS, PLA, HIPS, nylon, PVA, Laywood, and more.`
- ■ **OS supported** `Linux, Mac, Windows (MatterControl software is Windows-only.)`
- **Print untethered?** `No`
- ■ **Open-source hardware?** `No`
- ■ **Open-source software?** `MatterControl isn't, but OSS toolchains are listed in "downloads" on their website.`
- ■ **Printer control software** `MatterControl`
- **Slicing software** `Slic3r`

Taking cues from both the RepRap Air and Prusa, the Airwolf AW3D XL is a solid, robust machine with a lot of potential that produces quality, large-volume prints.

Although beginners might find this Rep-Rap-style machine a little daunting, Airwolf provides ample documentation and support from start to finish. An included USB drive contains sample STL files and Slic3r profiles. The manual guides first-time users through a beginner workflow with MatterControl, a free application available from MatterHackers (Windows-only at press time, Mac version promised). We actually found ourselves using this simple and effective software more than the Repetier/Slic3r combo.

Hardware-wise, the Airwolf has a heated glass bed that makes it suitable for printing a variety of materials. Its acrylic frame gives it both structural integrity and a graceful demeanor. The electronics and wiring are mostly kept out of sight and away from prying hands, making it suitable for most environments.

When we found ourselves challenged by a blocked extruder, Airwolf's customer support was awesome, getting back to us by phone and email within 20 minutes and guiding us through the process of locating and fixing the jam. However, reaching the source of the blockage was not as straight-forward as it could have been. Hopefully future iterations will address necessary access points.

PRIMO FEATURES
- ■ `Large glass heated bed allows for printing in ABS and will not warp.`
- `The Deluxe model includes interchangeable 0.35mm or 0.50mm nozzles.`

PRO TIPS
- ■ `If your nozzle jams, call Airwolf's fantastic customer support.`
- `Use the Windows-only MatterControl software to simplify your slicing experience.`

MBOT CUBE II

MAGICFIRM LLC / MBOT3D.COM

- ■ **Price as tested** `$1,399`
- **Print volume** `10¼"×9"×7¾"`
- ■ **Heated bed?** `No`
- ■ **Printer control software** `Replicator G`

This printer is an unfortunate example of how an open-source project can be exploited to produce overpriced, inferior hardware. Its sleek exterior belies shoddy workmanship and poor design — a kind of hasty mashup of MakerBot's Replicator 1 and 2 designs. To make it print ABS without a heated build platform, they've used a specially engineered plastic that hot ABS sticks to. Unfortunately it sticks too well, and much patience and/or uneasy amounts of force are required to remove a printed part.

—Tom Burtonwood and Derek Poarch

LEAPFROG CREATR

LEAPFROG BV / LPFRG.COM

- ■ **Price as tested** `$2,500 with VAT, $2,030 without`
- **Print volume** `9"×10.5"×7.8"`
- ■ **Heated bed?** `Yes`
- ■ **Printer control software** `Proprietary version of Repetier-Host`

At first the Creatr seemed great: It's big, solid, well-arranged, and sports an enclosed all-metal frame. Unfortunately we had trouble getting any prints longer than one hour to run successfully. We tried both extruders and had feeding issues with each. The documentation is great, but it's only for Windows. With tuning, it could be a solid unit if you've got the time and space for it.

—John Abella

SLICING AND CONTROL SOFTWARE
Where it's going and where it's been. *WRITTEN BY JOHN ABELLA*

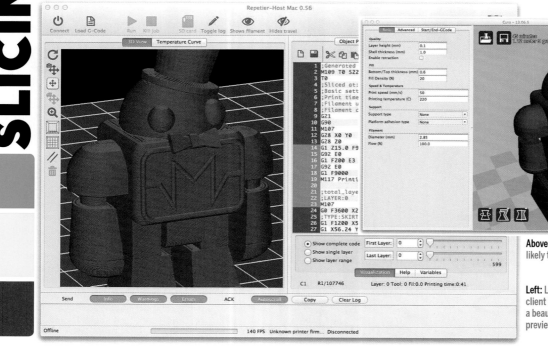

Above: Cura's slicing engine is most likely the fastest now available.

Left: Leading open-source printer client Repetier-Host sports a beautiful 3D pan/zoom/tilt preview window.

Models go through two software processes on their way to becoming finished prints: *slicing* and *sending*. Slicing divides a model into printable layers and plots the toolpaths to fill them in. The printer client then sends these movements to the hardware and provides a control interface for its other functions.

	Slic3r	Skeinforge	KISSlicer	Custom Open	Custom Proprietary
2013	53.3%	20%	3.3%	6.7%	16.7%
2014	55%	10%	10%	5%	20%

Among slicing engines used in our printer round-ups, the trend away from **Skeinforge** continues in 2014, and Alessandro Ranellucci's **Slic3r** continues to be the most popular slicing engine. Slic3r is open-source, cross-platform, and faster and easier to use than Skeinforge. Many vendors provided Slic3r INI files preconfigured for their machines.

Underdog **KISSlicer** saw increased use and discussion. Though closed-source and proprietary, KISSlicer is the work of one man — Californian Jonathan Dummer — and is available in a fully functional free version as well as a $42 "pro" tool that adds support for multiple extruders, auto-packing of parts, and other advanced features.

	Repetier-Host	PrintRun	ReplicatorG	Custom Open	Custom Proprietary
2013	18.3%	45.0%	1.7%	6.7%	28.3%
2014	32.5%	17.5%	5%	10%	35%

Among this year's printer clients, German offering **Repetier-Host** overtook Kliment Yanev's **Printrun** suite (best known by its GUI, Pronterface) as the most popular open-source choice. Printrun features an extensive macro language for command-line control, while Repetier-Host boasts a more graphical interface showing rotatable 3D views of models in the build area, plus a toolpath visualizer showing exactly what the printer will be doing during a build. Both programs smoothly integrate Slic3r to provide all-in-one printer frontends.

This integration of slicing engines into printer clients is an ongoing trend. Though open-source tools still dominate the scene, we also saw an increased incidence of closed, proprietary, all-in-one model slicing/sending programs from various manufacturers.

David Braam's **Cura** is an interesting exception. Though it's fully open source and can run many different printers, Braam works for Ultimaker, and Cura is largely associated with their brand. As of version 13.06, Cura introduced a custom slicing engine written in C++ that's probably the fastest available at the moment. In Cura, there is no slicing "button" — any time you make a change, the engine reslices automatically in the background. It usually takes only 5–10 seconds even on modest laptop hardware and looks, for the moment, like the shape of things to come. ✏

➕ For more software options, see page 34.

ENTER MICROSOFT

This summer, Microsoft announced that Windows 8.1, due out in October, will have native 3D printing support. The official Windows blog summarized their goal "to make 3D printing on Windows more like 2D printing on Windows." Whether these features land with a fanfare or a fizzle, no doubt the printer software landscape will look very different in 12 months.

LIQUID DREAMS

The desktop market for liquid-resin printers is small but full of potential.

WRITTEN BY ANDERSON TA, ERIC CHU *and* ANNA KAZIUNAS FRANCE

Liquid-resin printers boast faster print times at higher resolutions than common fused-filament printers, and they're higher-priced accordingly. Although stereolithography (SLA) is the original form of 3D printing, pioneered by Charles Hull in the 1980s, these desktop versions are brand new on the scene. They're called by different names, with laser-based printers going by SLA or SL and projector-based printers going by DLP (digital light processing). But the principle, in each case, is the same.

Instead of thermoplastic filament, these printers consume liquid photopolymers that solidify when exposed to UV light. To create a 3D print, the light draws a series of two-dimensional "slices" on the surface of a pool of resin, selectively curing it to form each cross-section of the object, layer by layer.

As a print progresses, the build platform slowly lifts the solidified plastic out of the liquid vat, exposing fresh layers of uncured resin at the interface. After each layer cures, the vat performs a "peel" movement to release the cured plastic and allow the fresh resin to flow in. This process repeats thousands of times during an average print.

When the print's done, the model is detached from the platform, excess resin is rinsed off with isopropyl alcohol, and support structures are removed with simple manual tools.

Currently, there are two main players in the desktop liquid printer marketplace: the laser-based Form 1 by Formlabs and the projector-based B9 Creator. We reviewed both. Read on to find out what we learned.

Jeffrey Braverman; 3D model by George Hart

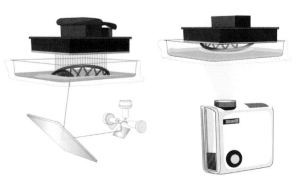

James Delaney/Formlabs Inc. (left); James Burke (right)

Laser-based SLA (or SL) printer (left), versus projector-based DLP design.

Why Choose a Liquid-Resin Printer?

PROS

- Higher resolution, smoother surfaces than fused-filament printing
- Faster — seconds per layer, not minutes
- More complex geometries possible
- Mechanically simpler
- Quieter

CONS

- More expensive hardware and consumables
- Limited selection of materials and colors
- Handling/disposal of liquid resin
- Cleaning uncured liquid from prints
- Prints are more brittle/fragile

FORMLABS FORM 1

WRITTEN BY
**ANDERSON TA,
ANNA KAZIUNAS FRANCE,**
and **ERIC CHU**

FORMLABS / FORMLABS.COM

- ▶ **Price as tested** $3,299 assembled, includes 1 liter resin
- ▶ **Build volume** 4.9"×4.9"×6.5"
- ▶ **Print materials** UV-cured resin
- ▶ **OS supported** Mac, Windows
- ▶ **Print untethered?** Computer can be disconnected after print begins
- ▶ **Open-source hardware?** No
- ▶ **Open-source software?** No
- ▶ **Printer control software** PreForm
- ▶ **Slicing software** PreForm

Gregory Hayes

The Form 1 is the first entrant into the consumer SLA 3D printing market, built by Cambridge, Mass.-based Formlabs. As of press time, it's also the highest-funded 3D printer on Kickstarter ever, raking in just under $3 million. The Form 1 uses a UV laser beam steered by a mirror galvanometer system to selectively cure a liquid photopolymer film to form each layer.

Sharp Looks, Seamless Function

This is a gorgeous machine — sleek and minimalistic with a matte aluminum chassis, OLED display, bright orange cover, and single-button interface. It has two cable connections: one power and one USB.

The Form 1 works as well as it looks. The software interface is streamlined to match the hardware; the process really is as simple as loading the file and hitting Print (or "Form," as they call it). The PreForm software also has a unique, built-in feature that notifies you if the model has errors and helps you orient

it correctly for printing. More experienced users can tweak options ranging from layer resolution to support structure parameters.

The instructions are simple and clear. Between them and the intuitive interface, we had our first object printing in less than 30 minutes out-of-the-box.

Forming Up

We tested the Form 1 using the bundled clear resin (an opaque grey is also available). The software offers three different resolutions: 100, 50, and 25 microns. The finer options dramatically increase print times, so we chose the lowest resolution (100 microns) in order to try as many prints as possible.

We ran prints of many sizes and durations; some filled the entire print volume and some ran well past 20 hours. The

> **Formlabs has created a modern marvel — a thoroughly enjoyable user experience and fast, high-quality prints.**

Form 1 never hiccupped. The parts were absolutely beautiful, the layers, even at the low setting, barely distinguishable.

We found the cleanup process a bit of work, as prints stuck really well to the platform and required considerable force to remove. Support structures printed under the default settings were also a bit harder to take off than seemed necessary. A small toolkit is bundled with the printer for this purpose, which is a nice touch.

All in all, Formlabs has created a modern marvel in its debut 3D printer — a thoroughly enjoyable user experience and fast, high-quality prints. ◾

PRIMO FEATURES
- ▪ Needs little maintenance, setup, or calibration
- ▪ Software automatically generates support structures and clearly marks problem areas on models.

WHO'S IT FOR?
Designers, Architects, Professionals, Makers

PRO TIPS
- ▪ The auto-orient feature positions the model for optimal curing and works great. Use it!
- ▪ Decrease the support structure's "touch point diameter" for easier removal and fewer surface blemishes.
- ▪ Decrease the base size; the default is thicker than needed.

HOW'D IT PRINT?

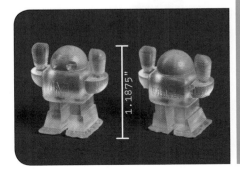

1.1875"

B9 CREATOR

WRITTEN BY ERIC CHU

B9 CREATIONS / B9CREATOR.COM

■ **Price as tested** $3,375
■ **Build volume** 4"×3"×8⅛"
■ **Print materials** UV-cured resin
● **OS supported** Linux, Mac, Windows
■ **Print untethered?** No
  Open-source hardware? Yes
● **Open-source software?** Yes
■ **Printer control software** B9 Creator Software
■ **Slicing software** B9 Creator Software

Gregory Hayes

With over $800,000 in funding from two Kickstarter campaigns, B9's Michael Joyce has created a well-built resin-based 3D printer using anodized aluminum parts and an off-the-shelf DLP projector. Providing exquisite detail capabilities, this open-source machine has been adopted largely by jewelers to print rings for lost-wax castings.

Manual Calibration

Setting up the Creator requires manual calibration, but the video walkthroughs of the hardware and software are very clear. You'll need to level the print bed with the silicone layer of the resin vat and focus the projector to your desired resolution. The software's bed leveling and projector calibration wizard walks you through these tasks. It took us about 15 minutes to dial it in at 50 microns, the B9's finest x-y resolution.

Powerful Software

The software gives you a lot of print setup options: You can manually attach support structures anywhere on the part, adjust layer exposure times, and tweak firmware settings to support different-sized projectors. The x-y build area varies according to the layer resolution you choose: 50 micron, 2"×1½"; 75 micron, 3"×2¼"; or 100 micron, 4"×3". The lack of automatic support generation is a drawback, especially if you're printing complex or overhanging shapes.

The Creator does not come with finishing tools, but it does have a parts list of gear to buy. We recommend getting a sealable tub to bathe your prints, like the one that comes with the Form 1, preferably large enough to soak the build table.

In testing, we used the 50.8-micron

> ### The B9 Creator produced the finest prints of any machine we tested.

layer setting. Layers of the print were still faintly visible on curved edges, but both vertical and horizontal flat sides looked completely smooth. The logos on the MAKE robot were very sharp, with only tiny steps on the tapered surfaces — still vastly better than an FFF machine. To further test fine detail, we printed a batch of robots at 0.35 scale; while the logos were light at that tiny size, most were still visible.

Conclusion

Capable of extremely small details, the B9 Creator produced the finest prints of any machine we tested. The software is fluid and responsive, with powerful configuration options. If you need professional-grade printing and are willing to spend time calibrating and experimenting, you'll be well-served by this machine. ◢

PRIMO FEATURES
■ Pro-grade DLP projector
■ Very high z-axis resolution, configurable from 6.35 to 101.6 microns

WHO'S IT FOR?
Makers, Tinkerers, Designers, Jewelers

PRO TIPS
■ The B9's software for manually adding support is slick, and you can use it for your FFF printer too. Export an STL and use NetFabb to trim any protrusions on the bottom; just set the position to 0mm and slice it flush.
■ Watch the Getting Started videos before setting up. The documentation isn't well organized.
■ The cherry resin can be tricky to use; for settings, check out B9 community member "CarterTG".

HOW'D IT PRINT?

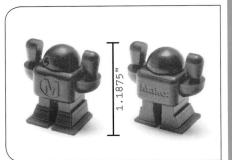

1.1875"

THE STANDOUTS

WRITTEN BY ANNA KAZIUNAS FRANCE and 3D PRINTING TESTING TEAM

THE BEST FROM OUR TESTS

BEST VALUE
Printrbot Simple
(page 60)
Runner-Up: Ditto+ (page 78)

BEST IN CLASS: PROSUMER FFF
Replicator 2 (page 68)
Runner-Up: Ultimaker 2 (page 66)

BEST IN CLASS: JUST HIT PRINT
Up Plus 2 (page 70)

SURPRISE HIT
Felix 2.0 (page 76)

BEST IN CLASS: RESIN
Form 1 (page 92)
Runner-Up: B9 Creator (page 93)

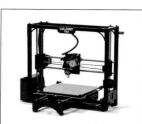

BEST DOCUMENTATION
Lulzbot TAZ (page 82)
Runners-Up: Replicator 2 (page 68),
Felix 2.0 (page 76), **Ditto+** (page 78),
Form 1 (page 92)

BEST OPEN-ARCHITECTURE
Ultimaker (page 66)
Runners-Up: Printrbot (pages 60, 86),
Deezmaker (pages 64, 84),
Lulzbot (page 82)

OVERVIEW: HOW THEY COMPARE

Through our testing, the machines began to organize into specific groupings. Here's the shakedown.

Ultra Compact

Trading volume for incredible portability, at a surprising price.

◾ **Printrbot Simple** ◾ **Bukito**

"Just Hit Print"

Don't need to print big and don't want to tweak print quality? These printers are for you.

◾ **Up Plus 2** ◾ **Afinia** ◾ **Cube 2** ◾ **Up Mini**

SLA/Resin Printers

Exceptionally detailed prints; build volume is somewhat limited. Artist's and designer's tools of choice.

◾ **Form 1** ◾ **B9 Creator**

Prosumer FFF

We found these machines to have professional reliability and robust software; well suited for engineers and design firms.

◾ **MakerBot Replicator 2** ◾ **Ultimaker 2**

Middle of the Road

Not standouts in our testing, but deserving consideration. We were unable to dial these in satisfactorily during testing but look forward to their continued improvement.

◾ **Bukobot 8 v2** ◾ **Lulzbot TAZ**

◾ **Printrbot Plus** ◾ **3Dprinter4U Builder**

Surprise Hits

These printers pleasantly impressed us. Their print quality separated them from the "Middle of the Road" crowd.

◾ **Felix 2.0** ◾ **Ditto+**

Experimental

These prototype models in our tests will be revised before they make their way out into the world.

◾ **Mini Kossel** ◾ **Type A 2014 Series 1**

◾ **Airwolf 3D** ◾ **Leapfrog Creatr**

◾ **Type A 2013 Series 1**

Buyer Beware

These printers did not perform well, arrived broken, or broke during the testing.

◾ **MendelMax 2.0** from Maker's Tool Works ◾ **Solidoodle 3rd Generation** ◾ **MBot Cube II**

Looking Forward: Our Recommendations to Printer Vendors

◾ Untethered printing is good; the more onboard controls for this, the better.

◻ PLA cooling fans work wonders. Having one on all PLA-optimized machines will be a great benefit to users.

◾ More super-compact, durable, portable printers: Rapid prototyping is much easier when you can bring your machine with you.

◾ Automating functions like platform and nozzle leveling go a long way to solving the issues faced by all 3D printer users, particularly beginners.

3D SCANNERS

Finally, consumer-grade scanners for your desktop!

WRITTEN BY **NICK PARKS**

3D scanning at home was nearly impossible without expensive pro equipment or DIY setups, until now. Consumer-grade scanners have arrived, based on two main depth-sensing technologies. Triangulating laser scanners project a laser line on the object, record it with a camera, and compute the triangulated position of the surface. Structured-light scanners project a pattern onto the object and calculate depth by measuring distortions in the pattern. We tested one of each to see how they work.

DAVID LASERSCANNER STARTER KIT

DAVID VISION SYSTEMS / DAVID-3D.COM

Jeffrey Braverman

- ■ **Price as tested** $675
- **Object scan size** 0.4"–15.7"
- **Accuracy** ~0.5% of object size
- ■ **Scan time** ~40sec
- ■ **OS supported** Windows only
- **Open source?** No

This kit from Germany's David Vision Systems may feel like less than $675 worth of components — a Logitech webcam, Joby mini tripod, handheld line laser, cardboard calibration backdrops, and scanning software on a flash drive — but they integrate to make a complete laser scanner. You place an object in front of the calibration boards and shine the laser on it from various angles. The webcam captures the reflected laser light, and the software assembles the 3D model.

The kit yielded fairly detailed scans; we captured the shape of a quarter, but not its surface features. We found setup and use to be tricky, due to limited instructions and unintuitive software that requires you to align the scans manually, a tedious and time-consuming job. There's no turntable, and dark objects scanned poorly, a problem common to laser scanners (some manufacturers recommend dusting the object with talcum or cornstarch).

DAVID SLS-1

DAVID VISION SYSTEMS / DAVID-3D.COM

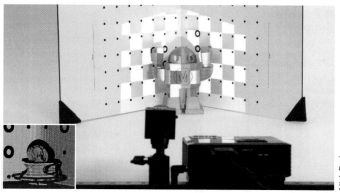

Nick Parks

- ■ **Price as tested** $2,700
- **Object scan size** 0.4"–23.6"
- **Accuracy** ~0.1% of object size
- ■ **Scan time** 2–4sec
- ■ **OS supported** Windows only
- **Open source?** No

David Vision's structured-light scanner, the SLS-1, uses high-quality components: an industrial-grade camera with focusable lens, a superb, focusable DLP video projector, a full-size tripod with an aluminum mount.

It projects black and white patterns on the object and scans them with the high-def camera, giving it much finer resolution than the laser kit — we captured Washington's profile and some lettering on the quarter — but it's also far more expensive.

Regrettably, the SLS-1 lacks a turntable, a key feature that makes 360° scanning easier and more precise. (DIYers have made Arduino-powered turntables for this scanner that reportedly work great.) Also, it uses the same tricky software as the laser kit, though the SLS-1's instructions are much better.

Overall, this is a solid 3D scanner for fast, accurate measurements. Its quality surpasses laser scanners, but at nearly twice the price of a MakerBot Digitizer, the lack of an automatic turntable is something buyers will have to weigh.

PREVIEW

3 HOT NEW SCANNERS

We're excited about three new 3D scanners that debuted just too late to review in this issue. All three have a turntable that rotates the object for complete 360° scanning, and all claim to be easy to use, with custom software that creates watertight scans ready for 3D printing. Watch for our complete reviews at makezine.com/review.
—Anna Kazuinas France

Spencer Higgins

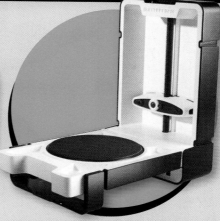

MAKERBOT DIGITIZER

STORE.MAKERBOT.COM/
DIGITIZER.HTML

- **Price** $1,400
- **Object scan size** 2"-8"×2"-8" dia.
- **Resolution** 0.5mm (0.0197")
- **Scan time** 12 minutes
- **OS supported** Linux, Mac, Windows
- **File formats exported** STL
- **Availability** Now

The MakerBot Digitizer uses a camera and dual lasers to scan objects, capturing details down to 0.5mm. It spins the object twice — once for each laser — then combines the point clouds in software and, for MakerBot printer users, exports the 3D mesh straight into MakerWare for printing.

MATTERFORM

MATTERFORM.NET/
SCANNER

- **Price** $579
- **Object scan size** Up to 9.8"×7" dia.
- **Resolution** Details to 0.43mm, size to ±0.25mm
- **Scan time** About 5 minutes
- **OS supported** Mac, Windows
- **File formats exported** STL, OBJ, and PLY (point cloud)
- **Availability** Expected late November 2013

The sleek, dual-laser Matterform looks similar to MakerBot's offering but at a much more maker-friendly price, and it promises the addition of full-color scanning, a moving camera head for capturing tricky angles, more file formats, and a clever portable case.

CADSCAN CUBIK

CAD-SCAN.CO.UK

- **Price** $1,125
- **Object scan size** 0.4"-7.8"×0.4"-8.6" dia.
- **Resolution** 0.05mm–0.15mm (50–150 microns)
- **Accuracy** ±50–150 microns
- **Scan time** 15 minutes+
- **OS supported** Linux, Mac, Windows
- **File formats exported** STL, OBJ, PLY, Sketchfab
- **Availability** Expected October 2013

CADScan's Cubik senses depth with a "series of phase-shifted patterns" projected by LED arrays and captured by dual 5MP cameras. It's their own patent-pending variant of structured-light scanning, and they're promising much higher resolution than laser scanners, with full-color capture as well.

UPGRADES, MODS, AND ACCESSORIES

Tools and modifications to help your printer work its very best.

Like cars and computers, 3D printers can be upgraded with all kinds of aftermarket goodies. We asked our testing team which ones they'd reach for first.

Upgrades

1. Heated print bed
A must for printing ABS, helpful for printing many other plastics.

2. Cooling fan
Machines that use PLA need a fan to cool the filament as it prints, greatly increasing print quality. Search Thingiverse for fan brackets and ducts to fit your printer. thingiverse.com

3. Glass build plate
Plastic build plates warp, an embarrassingly common problem in consumer-grade 3D printers. Glass won't.

4. High-temperature extruder
To print hotter-melting materials, try the single-piece, stainless steel Prusa nozzle, rated to 300°C. prusanozzle.org

5. OctoPrint host software
Print from anywhere on the planet and watch via webcam using this free web interface for (non-MakerBot) 3D printers. A special version runs on a Raspberry Pi. octoprint.org

6. Replicator 2 extruder upgrade
A far better filament tensioner, standard on newer MakerBots. Buy it or print your own. store.makerbot.com/extruder-upgrade.html thingiverse.com/thing:53125

7. Bottleworks upgrades
Beef up your MakerBot's Z-platform arms, and add a removable heated platform. bctechnologicalsolutions.com

8. Sailfish firmware
Unofficial MakerBot upgrade is a must for better printing; the community is a great resource for learning hardware, firmware, and software. thingiverse.com/thing:32084

Accessories

9. Palette knife
The old-school oil painting tool is a must-have for freeing objects "welded" to the build plate.

10. 3D Systems CubeStick glue
Useful for printing PLA in humid conditions, on any machine.

11. Rubber bulb air dust blower
To clean up your machine and your prints, it's cheaper and less wasteful than buying canned air.

Modifications

12. DIY auto-leveling probe
Forget manually leveling your build plate! Add a sensor that probes the bed height automatically. Search for "Z probe" at Thingiverse for various designs.

13. Stifle it
Modify your Afinia or Up printer to muffle the obnoxiously loud beeping. thingiverse.com/thing:31957

✚ For more great mods, check out these Thingiverse superstars: Dr. Henry Thomas (whpthomas) created a run of Rep2 mods including the popular extruder upgrade. Rich Cameron (whosawhatsis) opened eyes with his minimalist filament tensioner and other parts for various printers. Tony Buser (tbuser) is a go-to guy for printer mods, tools, and fun things to print. thingiverse.com

FILAMENT EXTRUDERS

WRITTEN BY
NICK PARKS
and
ERIC CHU

Filament is pricey — make your own and save with these extruding bots.

3D printers are great tools, but the cost of filament adds up fast. Filament extruders let you make your own by melting inexpensive plastic pellets. And you can experiment with different plastics and blends, to invent filaments with new colors and material properties.

We tested two machines; both are reliable and easy to use, with a push-button temperature controller and simple switches for the heater and motor. We tested only ABS extrusion, as PLA requires cooling and neither machine is equipped for it.

Ideally, filament extruders could also recycle your failed prints. These two can't, but the OmNom Project and Filabot's Re-claimer grinder (still in the works at press time) are aimed at this recycling problem, and we're watching with great interest.

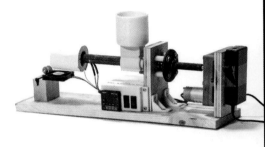

FILABOT WEE
FILABOT.COM
▶ **Price as tested** $749 assembled
▍ **Open source?** Yes

This is the Ferrari of filament extruders — it's fast, loud, and looks well-built. It can extrude 1kg of material in 5 hours. Even the fastest printers can't use up filament at that rate.

Unfortunately, like a Ferrari, it's also a bit expensive: $649 for the kit, $749 assembled. (It's an open hardware project, so you can also build it from scratch; see how on page 50.) Filabot also sells ABS pellets, priced from $5/lb (0.45kg) to $41/10lbs (4.5kg) for bulk quantities. Ultimately this machine is a good choice for people who want to make filament in bulk and maybe sell it, and don't mind a little bit of noise.

FILASTRUDER
FILASTRUDER.COM
▶ **Price as tested** $260 kit
▍ **Open source?** Yes

The Prius of extruders, the Filastruder is on the slower side, makes very little noise, and is well-priced — a bargain at $250 (or $260 with a 3D-printed hopper). Filastruder sells polyamide (nylon) powder, but recommends buying pellets from Open Source Printing: ABS for $8/2lb, PLA for $7/2lb.

The Filastruder takes 12 hours to extrude 1kg of filament, about 40% of the speed of the Filabot. It's as quiet as a 3D printer, emitting only a hum from its DC extruder motor. If you just want to make filament for yourself and a few friends, the Filastruder is an economical machine.

PRO TIPS

▶ Extrude from a tall table, at least 1m (39") high. On shorter tables, the filament may not have enough weight to pull itself downward and may build up a pile at the nozzle.

▶ Both machines can extrude 1.75mm or 3mm filament. Just heat up the nozzle, unscrew, and swap in the desired size.

▶ Filament winders are being developed (Filabot sells a beta kit) but manual winding isn't difficult. Don't wind it like a yo-yo since this introduces a twist that will cause tangles. Instead, spin the spool, using one hand to tension the filament as it wraps around.

▶ To make colored filament, mix special color pellets (known as "masterbatches") into natural ABS pellets at a ratio of 1:50, or 1:32 for richer colors.

▶ It's fun to create fades between colors. The effect gives printed objects an organic look.

Gregory Hayes

ONES TO WATCH

These promising new printers, scanners, and other 3D technologies have caught our eye.

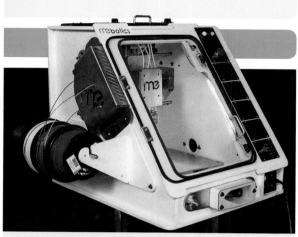

ALL-IN-ONE MACHINES

ZEUS
aiorobotics.com
- **PRICE** $1,999
- **BUILD VOLUME** 10.2"×7"×5.9"

The days of one-touch 3D copying may soon be here. Kickstarted and venture-funded AIO Robotics is targeting a casual and business audience with this networked, all-in-one machine that combines a 3D printer and 3D scanner to print, scan, copy, and even "fax" objects from one Zeus to another.

MICROFACTORY
mebotics.com/microfactory.html
- **PRICE** $4,495
- **BUILD VOLUME** 12"×12"×6"

Born in the Artisan's Asylum makerspace in Somerville, Mass., and billed as a "machine shop in a box," the Microfactory combines a 3D printer and a CNC mill for wood and plastics. With four print heads it can 3D-print in four colors or two different materials at once, and mill the results! It's run from an integrated onboard computer and can be used independently or networked. We wonder how well the vacuum port and filtered air inlet will keep dust from gumming up the printworks, but we're highly intrigued.

R.P.M. RAPID PROTOTYPING MILL
- **PRICE** $2,798
- **BUILD VOLUME** 12.6"×13.1"×10.8"

store.qu-bd.com

A large-volume 3D printer with CNC mill, the R.P.M. promises to print in ABS, PLA, and polycarbonate, and mill plastic, wood, steel, and aluminum. Though still in beta testing, R.P.M. also touts future interchangeable and add-on equipment (like a 3D scanner and stereolithography printing capabilities).

NEW FFF PRINTERS
Small and Cheap

THE BUCCANEER
pirate3d.com

- **PRICE** $347
- **BUILD VOLUME** 5.9"×3.9"×4.7"

Don't let the pirate theme fool you — the Buccaneer is all about sleek design and simple user experience at a category-killing price. Use cloud-based tools on your Android or iOS device to design, mod, and share objects, and print wirelessly on your desktop. The hype is huge, and the proprietary filament cartridges suggest they'll make their money on the consumables — but we're watching like everyone else.

PRINTRBOT GO V2
printrbot.com

- **PRICE** $1199 assembled
- **BUILD VOLUME** 8"×8"×8"

Literally a 3D printer in a suitcase, the Go is ultra-portable — it folds out when you want to print and tucks up neat and tidy when you're ready to go. Brook Drumm recently confirmed v2 is coming mid-October, with the same specs as the Plus.

ADVANCED FFF

HYREL ENGINE
hyrel3d.com

- **PRICE** $1,995-$3,095
- **BUILD VOLUME** 8"×8"×8"

Why stop with ABS and PLA? The Hyrel 3D prints in these plastics plus air-dry or plasticine clay, and even Play-Doh! Up to four print heads can be installed simultaneously and can be removed or replaced in less than a minute, even "hot swapped" at extruding temperatures.

FILAMENT EXTRUSION

EXTRUSIONBOT
extrusionbot.net

- **PRICE** $625
- **EXTRUSION RATE** 36ipm

Filament is expensive, so why not make your own? Pumping out 3–4 feet per minute, ExtrusionBot claims to be the world's fastest home-built extruder, and its handy spooling system keeps your filament from piling on the floor.

DELTABOTS RISING

"Delta robot" printers use a three-point vertical motion system to move the print head faster than Cartesian robots, and they can move equally fast in the x-, y-, and z-axis. Only one delta printer was available for MAKE's 3D Shootout this year (see page 62), but new ones are rushing to market.

DELTAMAKER
deltamaker.com

- **PRICE** $1,999
- **BUILD VOLUME** 11"×10" dia.

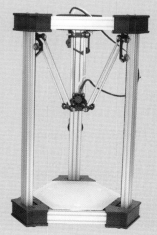

Kickstarted out of Orlando, Fla., the DeltaMaker uses MakerSlide aluminum extrusions paired with Delrin V-wheels on bearings to move fast and smooth. It's got an optional heated build plate and the largest build volume of the deltas we've seen.

Gregory Hayes

SPIDERBOT

spiderbot.eu

■ **PRICE** $1,140-$1,750
(kits only)
■ **BUILD VOLUME** 7"×7" dia.

With a precision aluminum frame, carbon fiber arms, and magnetically attached extruder plate, the SpiderBot is available as either a full kit (with optional full enclosure), or an upgrade kit to Spider-ize your existing printer.

ORION

shop.seemecnc.com

■ **PRICE** $1,499
□ **BUILD VOLUME**
8"×6" dia.

SeeMeCNC was the first to market with a deltabot kit, the Rostock Max. The Orion is their new fully assembled delta, with resolution as high as 0.05mm and a RAMBo controller board ready for hacks like adding lights or running more stepper motors.

SUMPOD DELTA

sumpod.com

■ **PRICE** $640
□ **BUILD VOLUME** 7.8"×7" dia.

Striving to bring down the price of 3D printers, this deltabot uses leadscrews for vertical motion and will accept an optional standard MK2 Prusa PCB heated bed. We're intrigued by the promise of an optional 3D scanning attachment.

REPRAP DELTABOTS

In addition to the Rostock and Kossel designs, check out these open source deltas in the RepRap community, not yet commercialized:

■ **Cerberus**
reprap.org/wiki/Cerberus, makerpair.com

■ **RepRap Simpson**
reprap.org/wiki/Simpson

■ **RepRap Morgan (SCARA ROBOT)**
Not actually a delta, but its hinged arm merits a look. reprap.org/wiki/RepRap_Morgan

NEW WAYS TO 3D PRINT

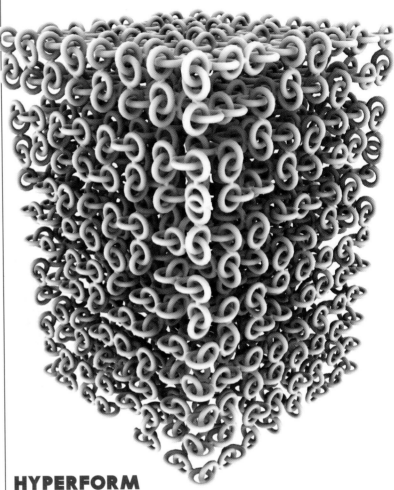

HYPERFORM

cmarcelo.com/#/hyperform

"4D printing" is shorthand for 3D printing objects that transform over time (the fourth dimension, see) to assume their intended shape. The idea has promise for printing and transporting things in small spaces, then unfolding them in big spaces (like space). MIT researchers Marcelo Coelho and Skylar Tibbits teamed with FormLabs to investigate folding strategies and print enormous chained objects in the comparatively small volume of the Form1 printer, and won themselves an Ars Electronica "The Next Idea" grant in the process.

DIWIRE BENDER

pensalabs.com

■ **PRICE** $3,000-$3,500
□ **BUILD VOLUME** Unlimited

This novel machine is a CNC wire bender, not a 3D printer, but it outputs your SVG files in perfect 2D curves of metal, to be assembled later into 3D creations. It handles wire up to 1/8" steel, bends 5 feet in 3 minutes, and looks beautiful doing it.

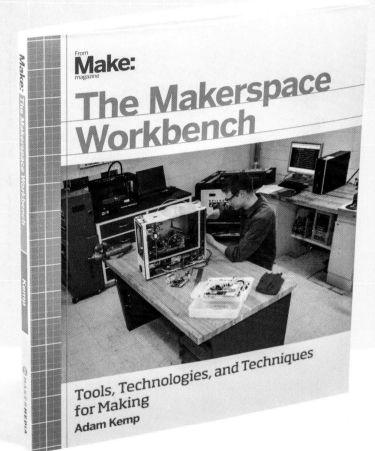

RESIN PRINTERS

NAUTILUS

indiegogo.com/projects/nautilus-3d-printer-dlp-tech

■ **PRICE** $387
□ **BUILD VOLUME** 4"×3"×4.7"

Built by makers in Beijing and aimed at a low price point, this DLP printer handles multiple resins and can hit 0.1mm to 0.01mm z-axis layer height. The kit is inexpensive compared to other printers, but you have to provide your own projector.

TRISTRAM BUDEL'S DLP PRINTER

instructables.com/id/DIY-high-resolution-3D-DLP-printer-3D-printer

■ **PRICE** $2,362
□ **BUILD VOLUME** 3.9"×3.9"×9"

Why buy a DLP printer when you can spend a year making one? Though still tweaking the design, Tristram Budel has put complete instructions for his scratch-built DLP printer on Instructables. At 1,000–1,500 man-hours, it's not for beginners.

MUVE 1

muve3d.net

■ **PRICE** $599–$1,099 kits
□ **BUILD VOLUME** 5.7"×5.7"×7.3"

With its variety of kit options and laser-cut wooden case, you could mistake the mUVe 1 for a throwback to FFF printers from 2008. Luckily, inside is a fresh resin-curing SLA printer, and its open-source design favors easily sourced, inexpensive parts ready for tinkering, like a 50mW UV laser upgradeable to 500mW.

OPEN3DLP

open3dlp.blogspot.com

■ **PRICE** $1,000
□ **BUILD VOLUME** 3"×4"×5.1"

Maryland Institute College of Art student and MAKE 3D Shootout tester Anderson Ta has developed a DLP printer from off-the-shelf parts. He's made some fantastic-looking prints and even detours into bioprinting.

MAKERJUICE

makerjuice.com

■ **PRICE** $40–$45/liter

MakerJuice's SubG resin comes in 8 different colors and cures with DLP projectors, UV lasers, and UV LEDs. They also sell pigments so you can mix the perfect resin color. Best of all, the resin is a lot less expensive than other suppliers.

3D SCANNERS

DIMBODY

indiegogo.com/projects/dimbody-3d-desktop-scanner

■ **PRICE** $540 kit, $810 assembled
□ **SCAN VOLUME** 12"×12"×12"

From Rimini, Italy, comes this open-source, Arduino-based 3D scanner that uses a rotating base, laser line, and monochromatic

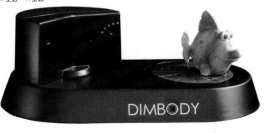

camera to build 3D models from real-life objects. It's accurate to about 400–100 microns and takes 8–24 minutes to scan an object, depending on resolution. We're intrigued by its Arduino roots; at press time it was an Indiegogo campaign in need of funds.

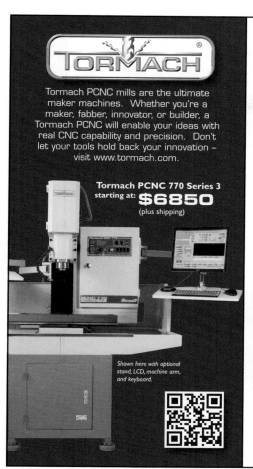

3D PRINTSHOP OF THE FUTURE

Get ready to print car parts, robots, wedding cakes, and even yourself. *WRITTEN BY* **TOM BURTONWOOD**

As 3D printers permeate homes and schools, the neighborhood 3D printshop can't be far behind. What will it look like?

Like a bike shop, you'll bring machines for **repair, upgrades, and calibration**. Like a skate shop, kids will congregate to hack DIY machines (and **print custom gear)**. "Internet café" variants will specialize in **3D printing foodstuffs.** Other functions I predict (some are already happening):

3D FEDEX/KINKO'S

Apartment buildings will have a printshop where residents can copy objects or pick up their brother-in-law "3D-faxed" to them (see page 100), using complex, multimaterial machines that don't fit in homes.

SPARE PARTS ON DEMAND

With the advent of precise metal printers, the spare parts economy moves entirely to local 3D printshops acting as hubs for Big Manufacturing — order an auto part on their app and pick it up an hour later.

3D PHOTO BOOTH

Photo albums are so 1989. Facebook is so 2009. The future is in 3D portraits (see page 44), our loved ones memorialized a la Han Solo

Nate Van Dyke

captured in time — first prom, football tryouts, ballet recital, wedding cake toppers, Christmas ornaments, Halloween masks ...

3D DESIGN SALON

Like beauty salons rent chairs, 3D printshops will rent stations where designers offer their services to repair broken parts, create new gizmos for the home, and work up Kickstarter mockups.

COMMUNITY PRINT-O-MAT

Hackerspaces will host laundromat-type lineups of 3D printers of all kinds, set up to run remotely or in person. Pay a fee to join the club and then pay by volume.

RENT A PRINTER

Test it for a weekend or lease it for a month. It works for Home Depot.

3D GEEK SQUAD

As printers become home appliances capable of printing more materials, they'll get bigger and less mobile. 3D printshops will send service teams on house calls to fix a kink in your feedline or calibrate your new firmware upgrades.

DIY PROSTHESES

Caregivers will scan people in need of prosthetic limbs, then print custom prostheses at the local printshop. (Robots will use the technology for custom repairs!) ◢

THE FUTURE IS NOW

Visit these 3D printshops today.

iMakr Store, London, opened April 2013
Inspired in part by last year's *MAKE Ultimate Guide to 3D Printing*, Sylvain Preumont launched the "world's largest 3D printing store," with retail sales, workshops, and print-on-demand service.

Deezmaker, Pasadena, Calif., opened September 2012
3D printer innovator Diego Porqueras (see page 25) launched this neighborhood hackerspace, 3D printer factory, and retail store.

The 3D Printer Experience, Chicago, opened April 2013
Artists and 3D print gurus helped launch this combo retail outlet, classroom, and design/print service — a project of The MetaSpace, a social entrepreneurship.

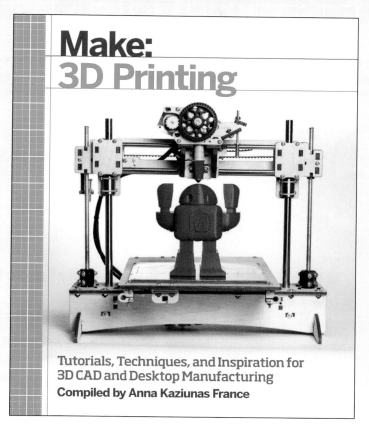

WRITTEN BY **CRAIG COUDEN**

DESKTOP MANUFACTURING IN THE MAKER SHED

Everything you need to start rapid prototyping at home.

One of the best things about hosting a 3D printer shootout is that after the testers go home, we in the Maker Shed have a ton of printers to play with! It's a great opportunity to evaluate hands-on what we want to offer you through **makershed.com**. After a lot of careful testing and review, here are the printers (and scanner) we currently stock.

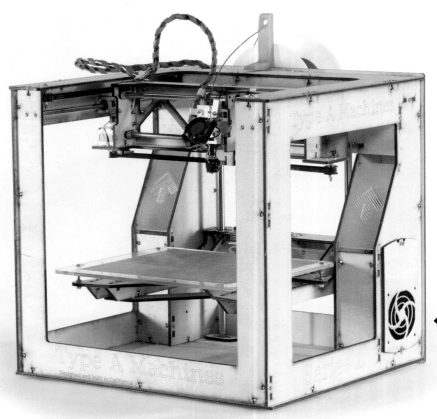

- **Afinia H-Series**
- **Cube**
- **Ditto**
- **Felix 2.0**
  ★ SURPRISE HIT 2014 ★
- **MakerBot Replicator 2**
  ★ PROSUMER BEST IN CLASS 2014 ★
- **MakerBot Replicator 2X**
- **Printrbot Simple**
  ★ BEST VALUE 2014 ★
- **Printrbot Jr V2**
- **Printrbot Plus V2**
- ← **Type A Series 1**
- **Ultimaker** ⎫
- **Ultimaker 2** ⎭ ★ BEST OPEN ARCHITECTURE 2014 ★

★ STANDOUTS IN THIS YEAR'S ROUNDUP

FILAMENT FOR DAYS

What next? It's time to stock up on filament! We carry both PLA and ABS filament in seven different colors (white, red, neon orange, natural, green, blue, and black). Available in 1.75mm and 3mm 1kg spools, our filament is made in the USA and has been rigorously tested by us to give the best quality prints possible.

Gunther Kirsch

MAKE: 3D PRINTING

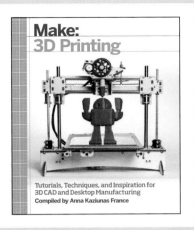

The perfect companion to our *Ultimate Guide to 3D Printing*, *Make: 3D Printing* is an essential book to keep within reach of your 3D printer. We've taken the best projects, tutorials, and stories about 3D printing from MAKE's print and online publications, freshened them up with the latest developments, and even added a few new pieces you haven't seen elsewhere. You'll learn how 3D printers work, what to expect in your first experience with one, and how to design, prepare, and print objects. You'll even learn finishing and repair techniques to make your objects look great. *Make: 3D Printing* makes it easy to understand the world of desktop manufacturing.

TINYG: MOTOR MASTERMIND

Riley Porter

Want to step it up? After watching a crazy-fast speed test video of a TinyG-driven Ultimaker (makezine.com/go/TinyG-test), we were hooked. TinyG has a four-axis stepper driver and an onboard microcontroller used to control 3D Printers, CNC machines, or anything that needs precise stepper control. The TinyG accepts G-code from your computer via USB and is completely integrated, so no worrying about individual stepper drivers, breakouts, shields, or annoying parallel ports. Few 3D printers can achieve TinyG's speeds (and precision) with the electronics they come with. Use the board to upgrade your current machine or to build one for yourself!

MAKERBOT DIGITIZER

Now you don't need any design, 3D modeling, or CAD expertise: The MakerBot Digitizer Desktop 3D Scanner scans objects up to 8" in diameter and 8" tall in about 12 minutes using a camera and two lasers. It outputs a standard STL file, which can be modified and improved in third-party 3D modeling programs. Just connect the Digitizer to a laptop or computer and you're ready to go. More specs on page 97.

Share your 3D printer with other makers in your own Makerspace!

Download the *How to Create a Makerspace* playbook!
http://makerspace.com/playbook

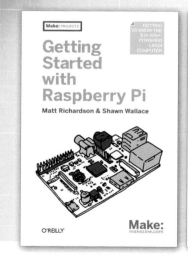
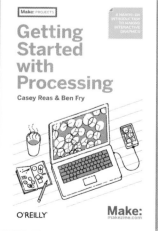
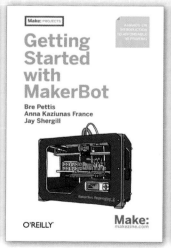
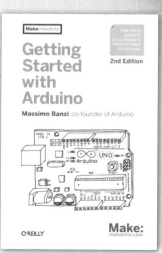

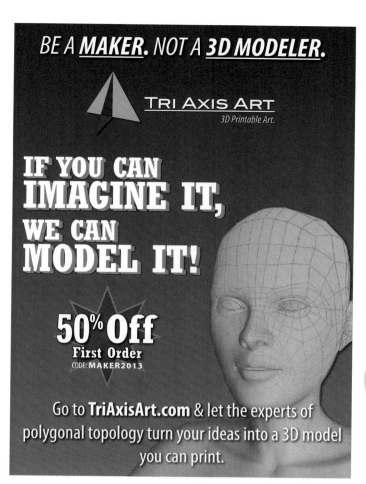

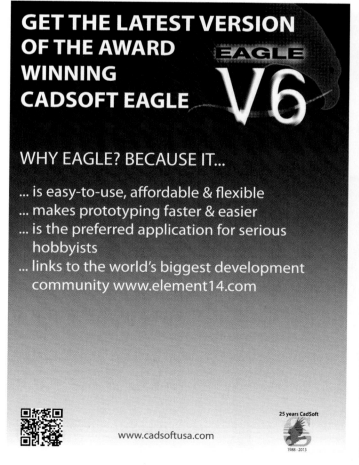

BUILD THIS!

Print your own articulating action-hero robot model.
WRITTEN BY **DAN SPANGLER**

MAKE YOUR OWN MEGA MAKE

■ Download the design files from makezine.com/go/mega-make

□ 3D Print

■ Assemble and glue

■ Save the world from alien invaders!

Mega Make's origins are a rough sketch in my maker's notebook, drawn in a sushi restaurant during lunch with my supervisor. The creation was to be the main protagonist in our stop-motion animated short about a giant robot fighting off an alien invasion. After much deliberation (and about a dozen California rolls) we finally agreed upon a design we both liked.

Using Autodesk Inventor's precise incremental dimensions and simple lines and curves, I recreated that drawing as a 2D digital sketch. I had the proportions, but our character also needed to move. I found some old patents online for G.I. Joe action figures that had

exploded views of various articulated joints that would work on the model. I also found various images of sci-fi robots to help inspire my design.

With the 2D sketch and reference material, I could start modeling the parts in 3D. I began with just the basic shapes of the torso, head, and various limbs, tweaking dimensions and contours to ensure the greatest range of motion for each piece. Once satisfied with the simple model, I could move on to my favorite part of the process: the details. I added bolts, vents, ridges, raised panels, rivets, power cells and more.

With the model finalized, I needed

to break its parts down into smaller chunks for 3D printing. I spent weeks printing out the parts and attempting to put them together, only to find that I fudged a tolerance somewhere and would need to reprint everything after adjusting the design's dimensions. Fortunately the 3D printer I use, an Up Plus, proved reliable and robust, letting me leave 12-hour prints running overnight without too much worry.

After about three weeks I had a finalized model hand-fitted and glued together. Mega Make now awaits his on-screen debut, full of moving parts and fun surprises, and I already have plans for the next one. ◪

Jeffrey Braverman